New Light on Therapeutic Energies

Mark L. Gallert

M.D., M.SC.

The Lutterworth Press

Cambridge

Published by
The Lutterworth Press
P.O. Box 60
Cambridge
CB1 2NT
England

e-mail: **publishing@lutterworth.com**
website: **http://www.lutterworth.com**

ISBN 0 7188 9008 6 hardback
ISBN 0 7188 9007 8 paperback

British Library Cataloguing in Publication Data:
A catalogue record is available from the British Library.

First published 1966 by James Clarke Co., Ltd
Reprinted 2002

TABLE OF CONTENTS

SOME ENERGIES GENERATED BY ARTIFICIAL CONTRIVANCES:

INTRODUCTION

THE ENERGY COMPLEX AS A FOCAL POINT FOR THE THERAPEUTIC APPROACH

To the extent that we can comprehend the difference between a living organism and that same organism after life has departed from it, we realize that the difference lies in the presence or absence of certain energy manifestations. Take, for example, a human body five minutes before death and five minutes after death. What is the difference? Size, weight and chemical constituents are virtually the same. Yet in the living body we have many manifestations of energy. To name a few:

    Flow of nerve currents
    Contraction of muscles
    Circulation of the blood
    Generation of heat
    Movement of food through the alimentary system
    Movement of liquid through the urinary system
    Movement of air through the respiratory system
    Production of matter to repair the various cellular structures.

None of these manifestations of energy are found in the dead body.

Except in cases of sudden death through accident, the transition from the living to the non-living state is simply the last step of a prolonged deterioration of the energy processes of the body. For example, as a person ages, the blood circulation diminishes, muscular contractions lessen in amplitude, until locomotion becomes more and more difficult—generation of heat lessens (old people feel colder than younger ones), production of material for cellular repairs becomes inadequate, so the organism deteriorates, and there is a corresponding diminution of every energy process in the body.

It would seem appropriate to express life in a quantitative as well as in an absolute manner—i.e., there is in general more life in a young person than in an older person, which is another way of saying that in a young person the energy processes take place at a higher level of amplitude.

One of the most common symptoms of illness is a feeling of weakness, which is simply a lessening of energy available for muscular contraction.

Ageing can be expressed as a progressive deterioration of the energy processes of a living organism, and illness can be expressed as a disturbance of the energy processes. The disturbance can be temporary, as in acute illness, or more lasting, as in chronic illness. Whether illness is the cause or the effect, of the energy disturbance, is a controversial point. . . . Death occurs with the cessation of the energy processes of the living organism.

In view of the foregoing, it is not surprising that there has been a great deal of research on therapeutic methods using various forms of energy.

Human thought tends to run in cycles or epochs, and we have been undergoing an epoch in which the main emphasis in therapeutics has been placed on the chemical aspects of the human organism—hence the wide use of drugs and the prolific discoveries and synthetization of new drugs. Yet, superimposed on what might be termed the chemical epoch, we can see at least the start of an energy one.

The concepts of scientific research as developed by the Western world in recent centuries, with the emphasis on established principles and the efforts to fit newly observed facts into those principles insofar as possible, has had the effect of encouraging the study and use of certain types of energy—namely, those with widespread uses in material or non-organic science, such as heat, steam, short-wave, etc. and has had the effect of discouraging the study and use of other types of energy. Physics has dealt mainly with non-organic uses of energy, and has attempted to explain organic application of energies in the light of the principles of non-organic uses of energy. It is our view that this has led to many misconceptions and blind spots in the study of the energy aspects of human and animal organisms.

It is becoming apparent, from research in various fields, some of which are outlined in this book, that the characteristics of the living organism embrace more types of energy than has previously been realized, and include some energy types that have not entered into the field of non-organic science.

Also, it is found that there are ways of applying conventional energies (i.e., energies widely used in non-organic sciences) to living organisms to produce effects that are not known in the non-organic or strictly material field.

It is a truism that the unconventional discoveries of today often become the accepted or conventional principles of tomorrow. However, this is not an automatic process. If discoveries of new and valuable facts are to be incorporated into the accepted scheme of things, they must be met with an open mind and adequate attention or consideration. Unfortunately these conditions frequently are lacking. There are many causes for this lack. Among these causes we may mention:

Economic restrictions
Psychological inertia
Vested interests
Institutional politics
And the general concepts of current scientific research previously mentioned.

These factors tend to favour the development and use of certain types of discoveries and to ignore or suppress the development and use of other types of discoveries. It is a purpose of the Foundation for the Study of Consciousness to do what it can to redress this balance. Toward that end, this book is issued, as an outline of a number of types of energies particularly applicable to organic science, and also contains outlines of relatively new applications of non-organic energies to the organic field.

In the search for means to combat illness and to prolong the useful and satisfactory portion of life, we can learn much from a deeper research into the energy complex which is an integral part of the living human body. We believe this book will show that the human energy complex contains elements which deserve more attention than they have received to date.

COSMO-ELECTRIC CULTURE

By George Starr White, M.D.

INDEX

COSMO-ELECTRIC CULTURE

THIS title, borrowed from George Starr White, M.D., includes six different types of energy arrangements having effects on plant, animal and human life that have been proven in practice. Since these effects are beneficial, it is hoped that further research will be done in these subjects in order to learn more about them. In this section we borrow liberally from Dr. White's book, *Cosmo-electric Culture*, 1940, with his kind permission.

1. EFFECTS OF METALS UPON PLANT GROWTH

When bright pieces of tin were dangled from fruit-tree branches by means of wires, it was found that not only were birds frightened away and the fruit thereby saved from destruction, but also that the fruit from branches thus protected was larger in size and of better quality than the fruit from branches that did not have the metallic application.

This is an example of energy from an inorganic substance, affecting the growth of a vegetable organism. This has been tried in many orchards, and with universally good effect.

To note the effect is one thing; to explain it is another matter. Is the benefit obtained from chemical absorption, or from some type of radiation emanating from the metal which is absorbed by the tree? Compare with the section on Eeman's relaxation circuits, where it was found that wires conveyed radiations from substances to living organisms, with a demonstrable effect upon the organisms.

Dr. White experimented with humus fertilizer stored in different types of containers. He found that plants fertilized with material kept in tin or iron containers had leaves of brighter green, and flowers with more vivid colours, as compared to plants fertilized with the same material kept in bags or in wooden containers. The same result was obtained by others who later duplicated the experiment, and by readers of newspaper articles reporting these findings.

2. EFFECTS OF GROUNDED METALS UPON PLANTS

A more pronounced effect than that outlined above, is obtained when the metal applied to trees or plants is grounded. Not only are the leaves and fruit in better condition, but scales and other parasites make little or no headway against trees with trunks protected by a circle of chicken wire extending about a foot below ground, and at least a few inches above ground.

Dr. White found that he could improve the growth rate, quality, and number of flowers on potted plants, by running copper wires from the earth in each pot, to a lightning-rod driven three feet or more into the ground. One ground rod will feed many pots. The contrast between the plants in the pots with grounded earth and those in pots left without the ground, was marked, indeed.

Dr. White's interpretation of this phenomenon is that there are certain energies in the earth which can be conveyed by metallic conductors to the plants, and will serve to strengthen the life energies of those plants. Since copper can form undesirable chemical combinations with the sap of plants and trees, it was found preferable to use iron wire or nails for making a specific contact with a tree, connecting the copper wire to the iron. Copper is the metal preferable for the ground connection, due to its resistance to rust.

From the earlier experiments mentioned here, a more elaborate layout was devised, termed "Cosmo-Electric Condensors", for stimulating the growth and improving the quality of trees.

This layout involves:

(a) Size and mesh openings of the rolls of wire placed around the trees.
(b) Spacing of the energy "condensors" from each other.
(c) Apparatus for energizing the irrigation water.
(d) Hook-up of iron-fence-posts to the trees.

Considering these in turn:

(a) It was found that the best coil for surrounding the trunk of a tree consisted of a roll of 24 in. wide, 1 in. mesh of chicken fence wire. It is interesting to note that a mesh less than 1 in. did not produce good results, nor did a mesh over 1 in. Apparently there is a resonance phenomenon here—certain beneficial frequencies for which the 1-in. mesh serves as a tuning circuit.

The 24-in. roll is then cut in two, and installed so that the cut edge is below the ground, and the selvage edge is 2 to 4 in. above the ground. The reason for this disposition of edges is not given, but the importance of following this instruction is stressed by Dr. White. The wire mesh is placed loosely about the tree trunk, allowing room for growth of the trunk. It appears to make little difference whether the mesh touches the trunk at any point, or is about an inch from it (see Fig. 1).

These and the following instructions are the results of many years of experiments conducted by Dr. White and others. Dr. White was in contact with thousands of people in all parts of the world, who used these methods. The fact that details are important, would seem to indicate that this type of method involves basic laws of considerable importance and power, but as yet we have only a faint glimmer of the import or character of those laws.

(b) The next point to consider, is that better results are obtained when trees are placed in regular rows than when they are spaced irregularly, also that the results are improved when the rows are lined up with the compass directions— i.e., North-South and East-West, and further it was found that a spacing of

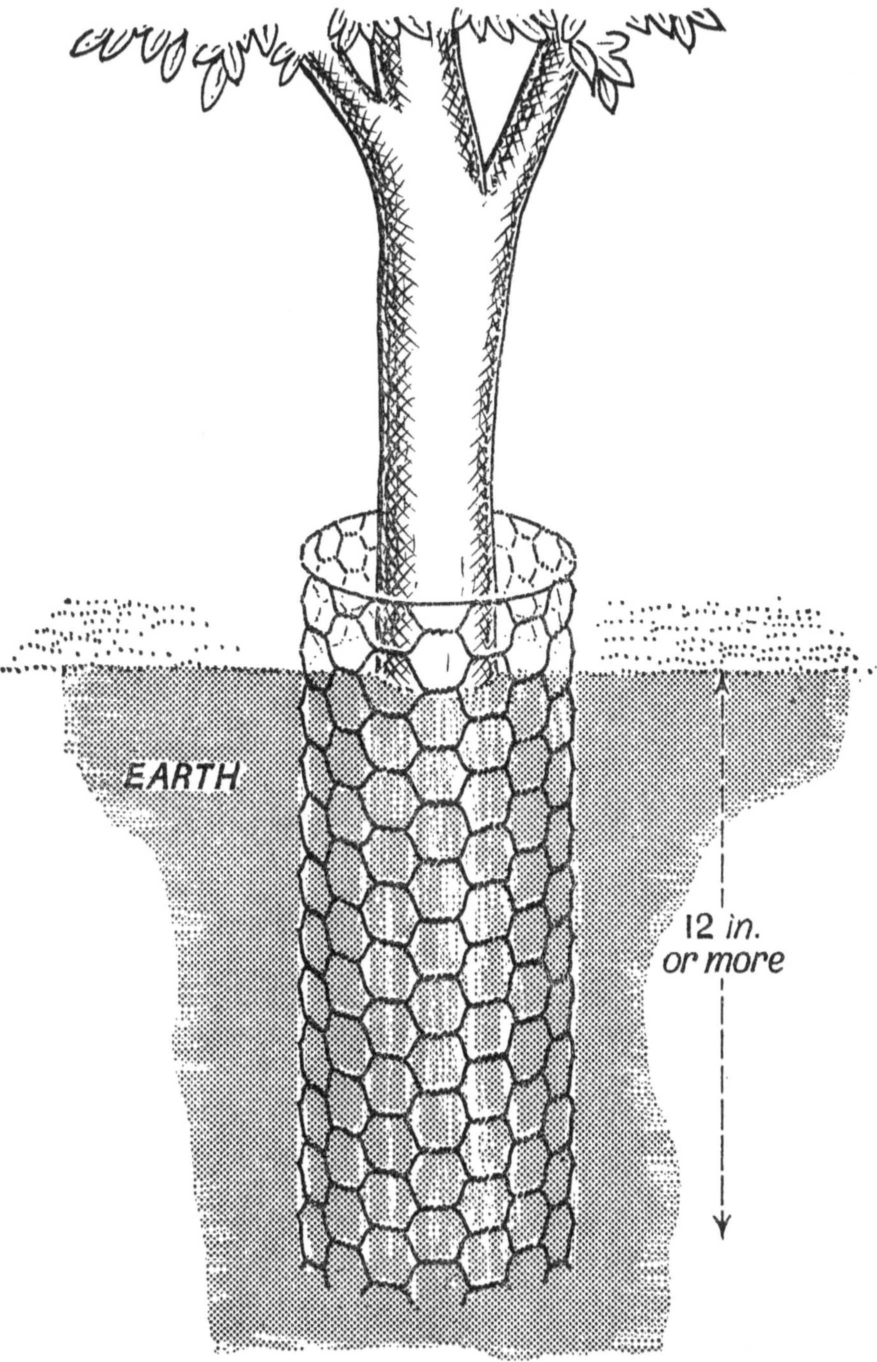

Fig. 1

30 ft.
30 ft.
30 ft.
30 ft.
A
A
A
A
A
A
Irrigation ditch
Irrigation ditch
2ft.
2ft.
2ft.
2ft.
30 ft.

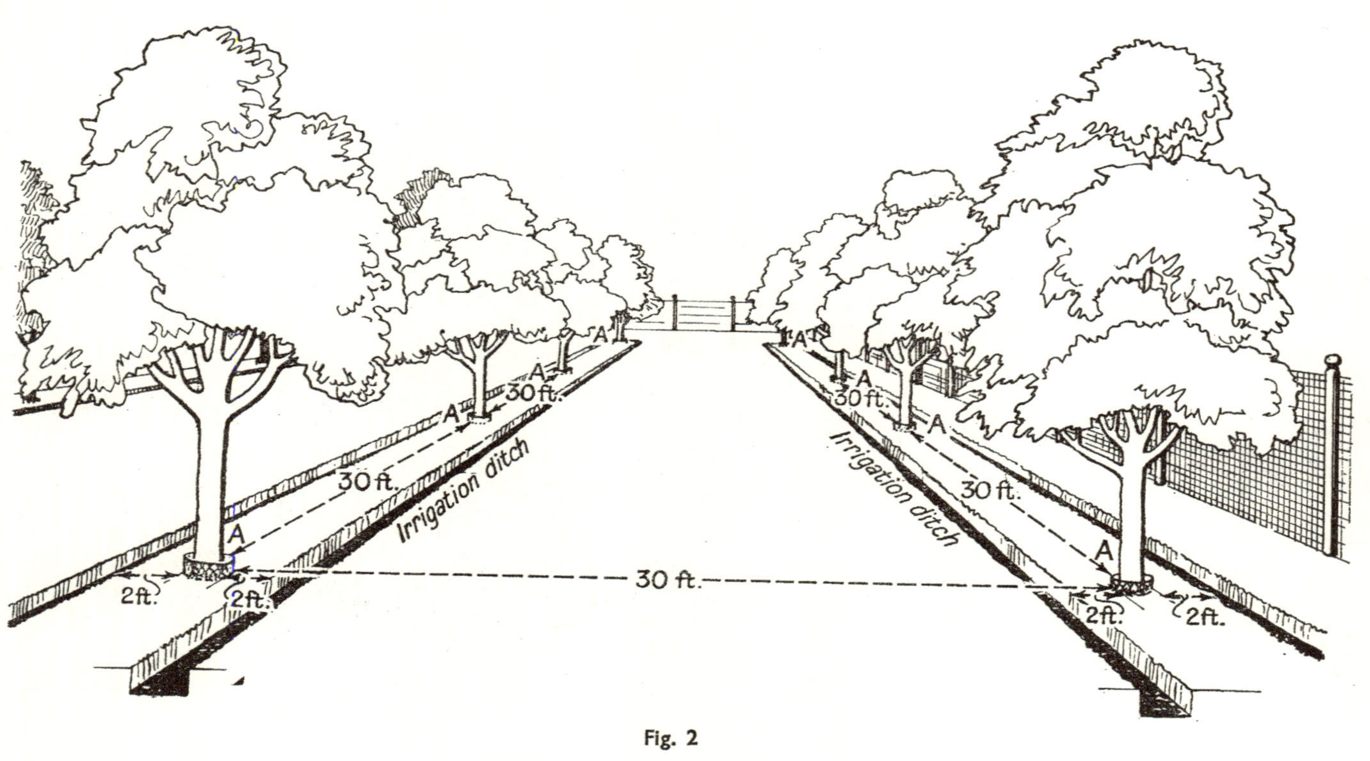

Fig. 2

thirty feet between trees was best, both between rows, and between the trees in each row. Irrigation ditches should be placed two feet from the tree trunks, on each side of each row (see Fig. 2).

(*c*) In the irrigation ditches, the water should be energized by the use of copper rods 3-ft. long at intervals of 10 to 20 ft. These rods are pointed at the ground end and with a copper or brass ball at the upper end. They are inserted in the middle of the irrigation ditch so that two feet of each rod is below the water level. Each rod is surrounded by a double coil of wire mesh, with the coil a little smaller in diameter than the width of the ditch. The mesh should be the same kind used around the tree trunks.

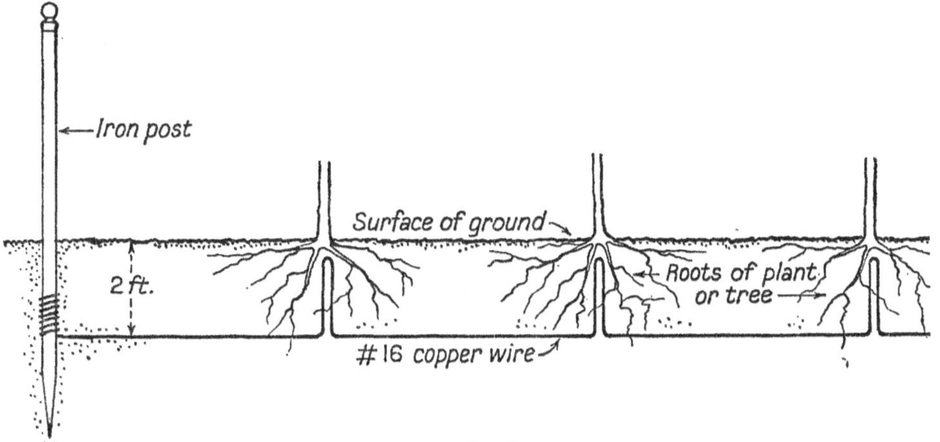

Fig. 3

(*d*) When possible, use an iron fence, or fence posts made of iron, and connect each post with copper wire #14 or larger, as in Fig. 3. The wire is led two feet under ground to a row of trees, with the wire extending up under the root system of each tree to a point near the ground surface level, then vertically downward to the 2-ft. level to the site of the next tree of the row, up under the roots to a point near the surface and down again, etc. until all trees in the row have been thus served by having the wire in proximity to the bottom of the tree trunk. The implications of these methods are quite interesting. About all we can do is to draw tentative hypotheses, since we lack instruments necessary to determine the exact manner in which these hook-ups work.

From section a., we have the effect of grounded metals. Besides the possible chemical effect from the contact of metal with the living vegetable organism, either directly or through the soil, there is the possibility of some type of molecular radiation from the metal. Why the grounded metal works better than ungrounded metal, for plants located in the earth and therefore already grounded, is not immediately clear. It may be that the metal serves as a distributor to convey to the tree some type of current or emanation from the ground, yet one would suppose that the tree was capable of absorbing such currents or emanations itself. At any rate there is obviously some type of inter-action between the ground and the metal which gives a more pronounced effect than is obtainable from the metal alone.

From section b., we find implications of spacing and direction. Why should trees spaced in regular rows do better than those irregularly spaced? Irregular spacing is a characteristic of nature—of uncultivated or forest lands. Is it that every organism has an effect on every other organism in its vicinity, and that an even line or row shares a "line of force", which enable the vegetable organisms to reinforce each other, so to speak? From the importance of aligning the rows in consonance with compass directions, there is the implication that magnetic lines of force from the earth may play a part in plant growth.

From the arrangement for "charging" or energizing the irrigation water outlined in c., it would appear that "something" is added to the water by metals imbedded in the earth, and that this enhances the qualities of the water in encouraging plant growth. We do not know if this effect is chemical, or electronic, or both.

The hook-up of the iron fence posts given in d., leads to still further implications. It would seem that whatever type of "energy" is drawn from the earth by the iron posts, can be conveyed to the trees by copper wire, and is absorbed by the trees if the wire is simply laid in proximity to the root system, without necessarily touching. Does the earth act as a "conductor" from the wire to the roots, or does the "energy" reach the roots by radiation? We do not know. Is the "energy" present in the earth and simply collected or drawn by the metallic fence-post, or is it the result of interaction between earth and certain metals? Again, at our present state of knowledge we can only guess.

3. EFFECTS OF GROUNDING ANIMALS AND HUMANS

(a) *Grounding of Animals*

After learning through observation that lice did not accumulate on roosts in hen-houses that had a wire connected to the ground, in contrast to troublesome accumulations of lice on roosts that were not grounded, many hen-houses were grounded, with consequent freedom from parasites and improved health of the poultry.

Dr. White then reasoned: "I observed that all vegetation was a part of the earth from which it grew; that animal life in the water was grounded to the earth through the water; that animals of the air were grounded when not flying, because the trees or other sleeping places they occupied, were attached to the ground; that animals living on the earth were naturally, in some manner, grounded; that natural animals were made un-natural by humans; that humans in their natural (uncivilized) state live grounded; that the further humans

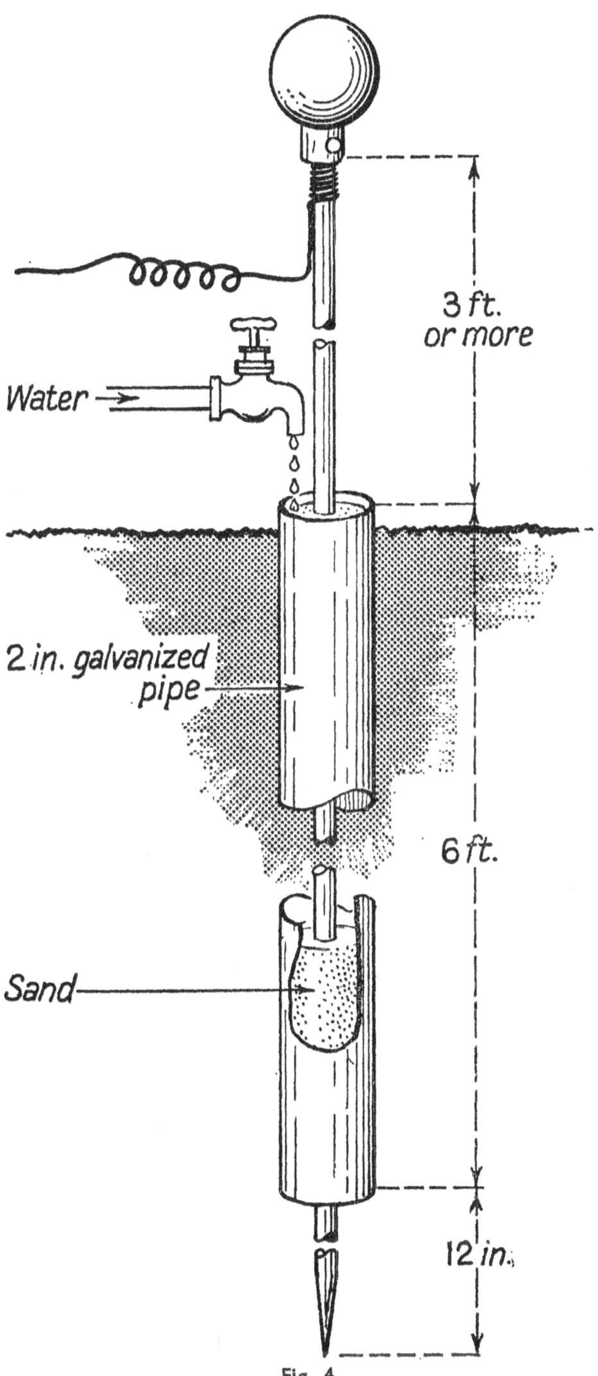

Fig. 4

departed from natural living the less they lived grounded and the more prone they were to illness."

(b) Grounding of Humans

This was the next step in Dr. White's progress into the use of natural methods, and not only proved to be of value in treating illness, but also in building immunity against infectious diseases.

(c) Methods of Grounding

1. To ground beds, place one end of a small copper wire under the bed pad; in a position under the middle of the back of the sleeper, solder the other end of this wire to a copper rod a half-inch or larger in diameter, driven five feet or more into the earth.

2. A stronger healing effect is gained by grounding with a device which is one form of what Dr. White terms a "Cosmo-Electric Energy Collector" (see Fig. 4). A 7-ft. hole is bored into the ground, to take a 2-in. galvanized iron pipe, which is inserted so that 2 in. projects above the ground. Placed centrally into the sunken pipe is a $\frac{1}{2}$-in. solid copper rod, 10 ft. long. Three feet of the length of this rod is left projecting above the galvanized pipe. The entering end of the copper rod is tapered so that it will go into the ground below the lower end of the galvanized pipe. With the copper rod held in position, regular mortar sand is poured in and gently pressed down. A pet-cock or small spigot is placed directly over the sand in the 2-in. iron pipe, so that water may be allowed to continually drip into the sand and thus keep the $\frac{1}{2}$-in. copper rod wet all the way down "and beyond". A solid copper, or brass ball 2 in. or more in diameter, should be placed on top the copper rod projecting 3 ft. or more above the iron pipe. This copper or brass ball should be bored so it will fit over the projecting copper rod and be securely pinned on by means of a collar on the ball drilled through and made secure with a brass pin. Such a ball should surmount every "grounding rod" and be thus secured, so that expansion and contraction from heat and cold cannot loosen it. A suitable copper or brass wire, No. #10 or larger, should be wound about the projecting rod, just below the ball, several turns and then soldered to the rod. Run the copper wire down the rod and pass it through a galvanized iron pipe to the house. Bore into the house and insert a regular electric-wire-insulator tube. Pass the grounding wire through the insulator tube. Inside the house, fasten some sort of clip which can be used for attaching the other wires.

It is specifically stated that the wire should not come in contact with the skin when sleeping. Enough energy will go through the mattress pad and under-sheet to energize the occupant of the bed.

(d) Treating with Cosmo-electric Energy

For relaxation or therapeutic purposes, place a copper hook over the wrist, connected to the Energy Collector device described in paragraph c-2 above.

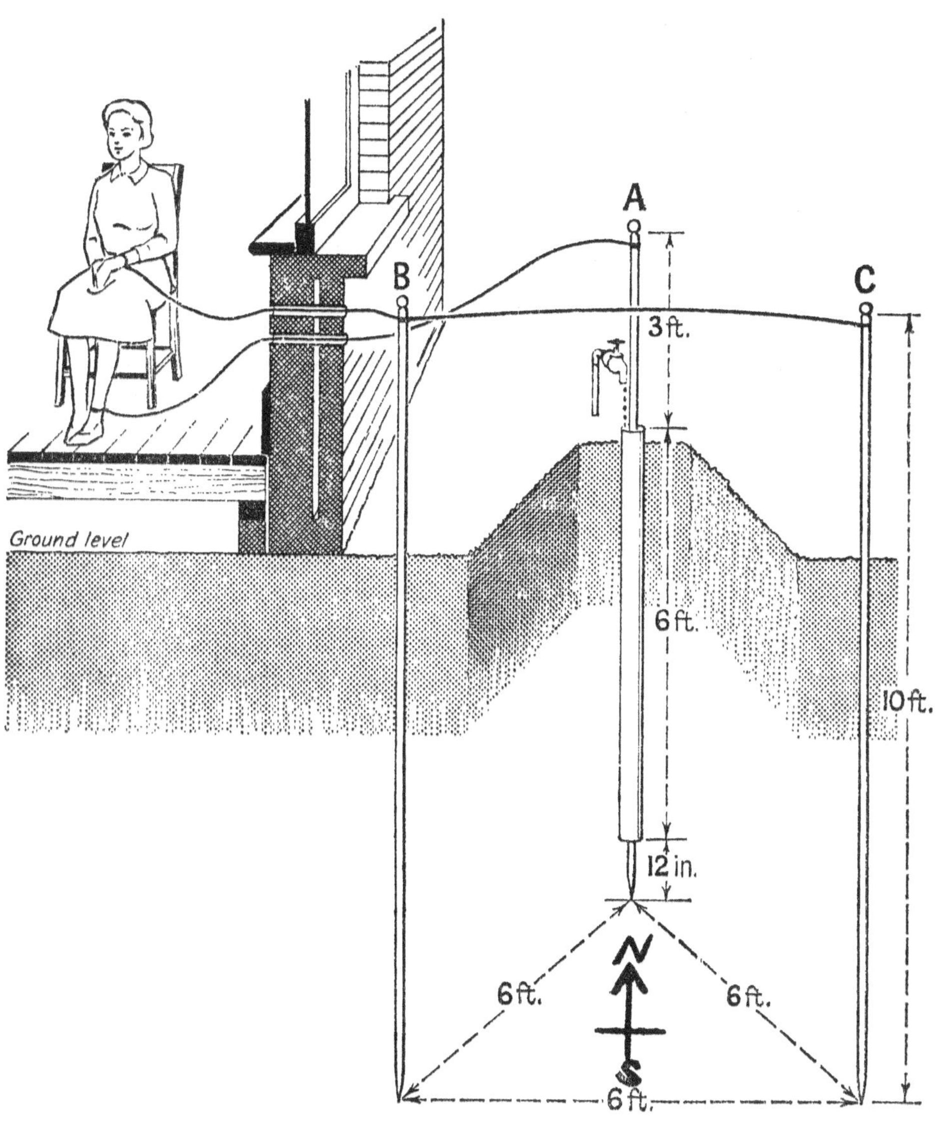

Fig. 5

11

One can benefit from this energy when reading, sewing, or engaged in any occupation while sitting down.

(e) Description of a Much Stronger Collector of Cosmo-electric Energy (see Fig. 5)

Three grounding rods are placed in the ground, spaced so as to form an equilateral triangle 6 ft. on each side. All are solid copper rods, at least 10 ft. long. Rod A is encased in a 2-in. galvanized iron pipe driven 6 ft. into the ground, with the copper rod extending below the bottom of the galvanized pipe at least 1 ft., and above the ground for 3 ft. Rods B and C are copper, of the same diameter as A, but driven into the ground without any casings. Rod A is the only one to be moistened artificially. The faucet that wets the sand about the Rod A, should be attached to a water pipe at least 6 ft. distant. The arm to which the faucet is attached should be about 1 ft. above the ground. Each of the three copper rods should be at least 6 ft. from any water pipe or gas pipe. Each of the rods is topped with a ball as described previously. Rods B and C are connected with a copper wire, soldered to each rod just below the ball. In use, connections are made to rods A and B (two separate leads) and taken into the house through insulator tubes.

The lead from rod A goes to one's ankle, from rod B to one's wrist, for a full-strength revitalization of cosmo-electric energy. Dr. White states that the energy concentrated in this device is equal to that obtainable from sixty miles of condenser wire surface.

Twenty minutes has been stated as a desirable duration for treatment. Hooks, bands or clips of copper can be used to make the contact between the lead wires and the individual.

Dr. White states: "The beneficial results (of this treatment) are unbelievable to any person not accustomed to see the invisible made visible—un-health changed into health—discord changed into harmony".

4. DIRECTIONAL EFFECTS

(a) Upon Animals

Observation showed that the egg production from poultry housed in roosts placed so that the hens faced North or South, was consistently larger than from the hens facing other directions, also that milk production was higher in cattle kept in stalls facing North or South.

Dr. White concluded that the same energy that makes the compass needle turn to a particular direction, has a great deal to do with life and the processes that make life possible upon this earth.

(b) Upon Humans

He found that persons with high blood-pressure or signs of hardening of the blood vessels, should sleep with the head directed East or West. Persons with low blood-pressure, or with a lowered vitality, should sleep with the head directed either North or South, and when taking a treatment of cosmo-electric energy as described in Section 3, should sit facing North or South during the treatment. Writes Dr. White:

"When we realize that our bodies are made up of 'polarized entities', and that the earth is a great magnet revolving in a universe of polarized energies, then we shall understand that the more we are in harmony with these COSMIC energies, the more benefit Nature will bestow upon us."

5. CERTAIN USES OF SUNLIGHT

(a) *To Charge Fertilizer*

Dr. White describes a container for compost and humus, made of wire screen mesh, thus charging the fertilizer not only with the metallic effects previously outlined, but also permitting sunshine to act upon the fertilizer with its consequent radiational charges (see Fig. 6).

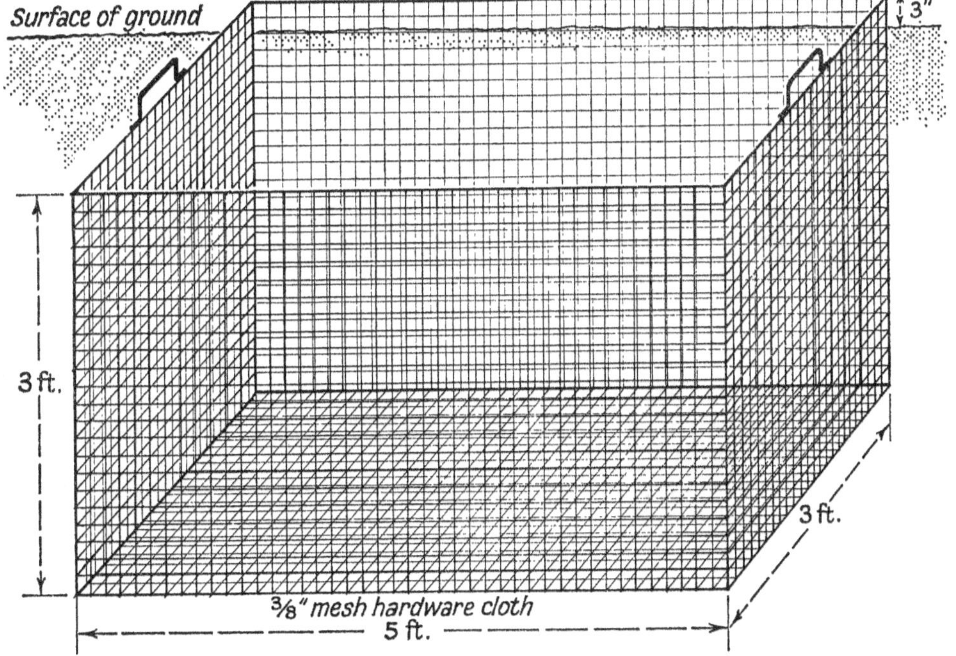

Fig. 6

(b) To Charge Foods

Many root foods, fruits, etc., are kept in the dark during the long winter months, peeled and cut for eating, then placed in the sunshine, outside of glass, for a few minutes. While this is done, they should be on earthen plates, or on any insulating material, rather than allowed to rest on any metal.

To explain the reason for this, it is stated that all cellular activities are the result of electrical charges. Sunshine on cells that have been in the dark for some time, starts up cellular activity, through the impact of magnetic or electrical energies. If this activity takes place on a metal plate, much of the "new life" is carried off into the atmosphere, but if on an insulating material such as glass, porcelain, or baked earthenware, the newly energized cells retain their magnetic or electric potency, according to Dr. White's theory. He states: "Eating 'electrically potentized' food is putting 'new life' into the system."

6. EFFECTS OF CHARGING WATER WITH ELECTRIC SPARKS

(a) For Plants

It was found that the use of irrigation water that had been ozonized, had a more stimulating effect upon plants than the use of ordinary water, but a stronger effect was obtained by sparking the water with electricity. The use of the 100-volt electric line for this purpose was found to be dangerous. The practical and safe way to spark water is to use a magnetic coil and dry batteries, as shown in the diagrams (Figs. 7 and 8).

In using a sparking device, the water should be in glass containers, or pottery ware. Put the ground end of the Sparker into the ground. Take the high-voltage conductor in one hand and press the switch button to turn on the device; at the same time bring the sparking terminal slowly over the water to be sparked. A fat spark will jump to the surface of the water. This spark will stop the spark that may show in the safety-spark gap. After a little practice one will get the spark on the water *before* the safety spark gaps. There are three ways to supply the sparking energy to vegetation:

(1.) Water the plants with sparked water, pouring it out of glass or porcelain or enamel vessels.

(2.) Spark the wire netting surrounding the trees, vines or shrubs as described in Section 1.

(3.) Spark the plants directly, by placing the ground terminal of the sparker into the ground and directing the sparks to the roots or crown of the object to be sparked. Two or three sparks directed to the crown of a tree each day will so improve the fruit, that only persons who have tried it for a season will believe that such improvements came from just sparking the tree.

It is stated that all of these sparking methods enhance the growth and healthfulness of the products produced from sparked vegetation.

High-voltage spark
Model "T" FORD spark-coil, or any six-volt vibrator-type coil
Vibrator
Switch
Four standard heavy-duty dry cells
Terminal for ground connection

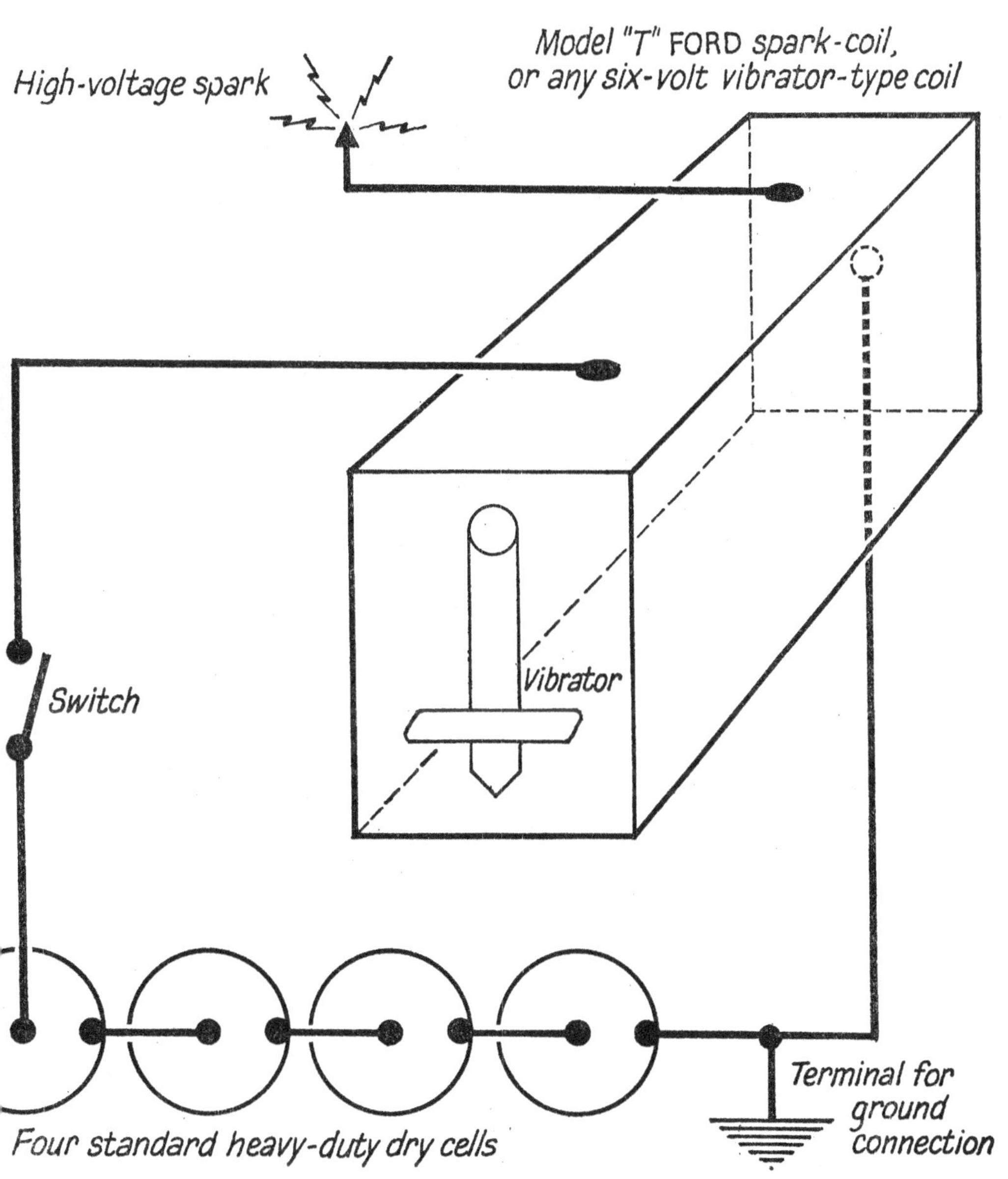

Fig. 7

15

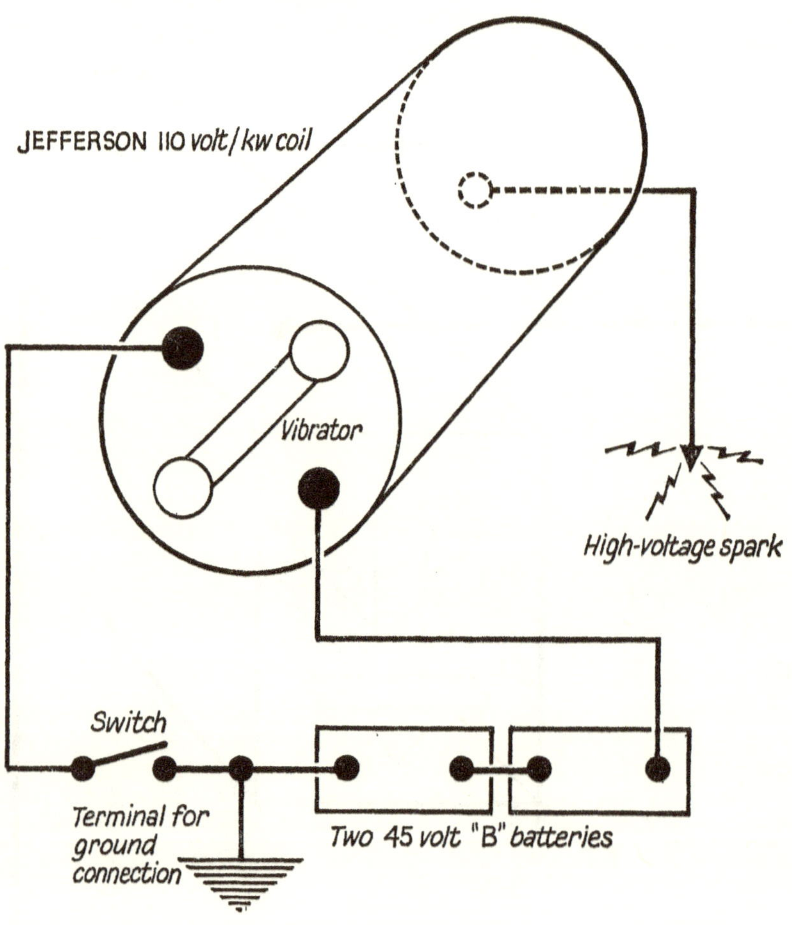

Fig. 8

(b) *For Humans*

The use of sparked drinking water is advocated by the author of "Cosmo-Electric Culture".

The sparked water will hold its energy for several hours. Dr. White states that it is well to use fresh drinking water every morning and spark it for use during the day as required. Be sure to use an insulated cup (one that is not made of metal). One touch with the body, or anything that is grounded, will discharge the energy from the water.

CHROMO-THERAPY

CHROMO-THERAPY

INTRODUCTION

ONE of the most interesting therapeutic methods based on vibratory phenomena is Chromo-therapy, or healing by means of beams of coloured light directed towards the body.

(a) *In modern times, interest in colour therapy was stimulated by the proven effects of different colours of light upon the growth of plants.* The first important book on the effects of different colours of light upon living organisms, was *Principles of Light and Colour* published in 1878 by Dr. Edwin Dwight Babbitt, M.D., of New York City—a classic on the subject, which examines many different aspects.

Dr. Babbitt's book includes references to Dr. Newberry's experiments in growing plants of the same variety, under white, red, and blue glass. Reference is also made to a previous writer—Robert Hunt, whose book *Researches on Light in its Chemical Relations* described the results of many experiments in growing plants under different colours of light.

Germination of plants was found to be speedier under the influence of blue light, often being accomplished in a period of two to five days instead of eight to fourteen days with the plant varieties under test.

After the seeds had germinated, it was found that growth above ground requires colours from both sides of the spectrum of visible light, but a *predominance* of yellow produced the best growth.

Flowering, seeding and fruiting were accomplished more readily by red rays. However, if plants are given too much red, they tend to wilt or wither. Plants that had been over-dosed with red light, were revived by applying blue and violet rays.

(b) *The first book written on the use of colour for therapeutic purposes*, was published in 1877 by Dr. S. Pancoast, entitled *Blue and Red Light*, or *Light and Its Rays as Medicine*. It dealt with the contrasting effects on the human body of the warm, stimulating red ray and the cool, soothing blue ray. Edwin Dwight Babbitt's more comprehensive book followed in 1878, containing a great deal of detailed information on the effect of several colours of the spectrum, and described their use in different ailments. This information though it proved of value to many doctors who adopted chromo-therapy as one of their methods of treatment, was based more upon empirical results than upon any basic understanding of the fundamental principles involved.

(c) *It remained for a Hindu scientist by the name of Dinshah P. Ghadiali to discover the scientific principles* which explain why and how rays of different colours have different therapeutic effects upon the body organism. These discoveries made possible a much more comprehensive and systematic method of colour therapy. Ghadiali came to the United States around the turn of the century, and after decades of research he published in 1933 the *Spectro-Chrome-Metry Encyclopaedia* in three volumes—a master work on colour therapy, though written in relatively non-technical language. It is now out of print and extremely rare. Ghadiali taught colour therapy to thousands, including many doctors, and developed improved types of colour-lamp equipment on which he obtained U.S. patents, that have since expired.

The material to follow in this section on chromo-therapy, is summarized mainly from Ghadiali's three-volume encyclopaedia on the subject. We shall start with a synopsis of the theory, which shows the scientific principles that underlie the proper use of colours for therapeutic purposes, and then proceed to the application of the theoretical principles in practice, as expounded by Ghadiali. The subject of Chromo-therapy involves a philosophy, a science, and a technique of application.

SYNOPSIS OF THE BASIC HYPOTHESES OF THE SCIENCE OF CHROMO-THERAPY

1. *Colours represent chemical potencies* in higher octaves of vibration. The relation between colours and chemicals will be described, also the advantages of applying colour vibrations of chemicals rather than to ingest chemicals into the body in material form as is done with the administration of drugs.

2. For each organ and system of the body there is a particular colour that stimulates and another colour that inhibits the action of that organ or system.

3. By knowing the action of the different colours upon the different organs and systems of the body as mentioned above, one can apply the correct colour that will tend to normalize the action of any organ or system that has become abnormal in its functioning or condition.

4. The process of life involves a proper balance within the body of all the different colour energies. When this balance is disturbed, disease results, and if the unbalance becomes too great, premature death ensues. The aim of the science of colour therapy is to combat disease by restoring the normal balance of colour energies within the body.

5. All elements known on earth are found in the sun, as shown by spectroscopic analysis. The sun's rays bring us the energies of every known element, from which all chemical combinations are made. White Light, the summation of all colours, contains the energies of all elements and chemicals found in the sun, both known and unknown. The sun is constantly pouring White Light energy into the atmosphere of our earth, thus "charging" this atmosphere with the different types of energy needed to sustain life.

6. Man has, surrounding and inter-penetrating his physical body, an auric body, which has as one of its functions the power to absorb the white-light energy from the atmosphere, split it into its component colour energies, and these colour energies then flow to different parts of the physical body to energize it. (Note: The auric body or human aura, formerly only seen by clairvoyants or certain psychics, was made visible to nearly everyone by equipment developed by the British scientist Kilner, and others, as will be outlined later.) The effects noted from the use of colour therapy are deemed to occur through the action of colour rays upon the auric body, which in turn influences the physical body.

7. In using colour-therapy in a scientific manner, one is making constructive use of the principle of "affinity"—one of the most powerful principles of chemistry and atomic physics, which governs the combining of elements into molecules. How the principle of affinity operates in colour-therapy will be outlined.

Chapter 1 COLOURS REPRESENT CHEMICAL POTENCIES IN VIBRATORY FORM

A necessary step in understanding this point is to obtain a grasp of the spectrum of light, and the science of spectrum analysis.

A beam of light from the sun, passed through a prism shaped in the form of an equilateral triangle, emerges in a seven-band spectrum having segments of violet, indigo, blue, green, yellow, orange, and red.

These seven basic colours are always present in the spectrum in the same order, identical with the order of colours in the rainbow ... the colours always are in the same respective position in relation to the prism, red appearing at the thin edge and violet at the thick end of the prism.

By using various devices incorporating magnification, the spectrum has been spread over a larger visual area for more minute observation.

With these devices, dark lines were discovered in certain portions of the solar spectrum, in 1802 by Dr. William Hyde Wollaston, an English doctor and chemist.

Joseph von Fraunhofer, a German optician and physicist, found in 1815 that the relative positions of these dark lines in the spectrum were always the same. He found that the burning flame of sodium gave a dark line in the spectrum produced by the flame, identical in spectral position with a dark line observed in the spectrum from the sun. Further research in 1850–59 by the German scientists Gustav R. Kirchoff and Robert W. von Bunsen resulted in the discovery of numerous other correlations between lines in the spectrum and identical lines from chemical elements burned in the Bunsen burner (gas flame).

It was later found that lines in spectra from electric sparks emitted by electrodes made of metals under test, and from sparks in tubes with elements in gaseous condition, are identical with the lines originating from same elements disintegrated by combustion (see Appendix at end of this section).

It was also seen that the colour of the flame produced by burning a substance or gas, depended upon the element or elements being burned. Carbon burns with a yellow flame, hydrogen with a red flame, oxygen with a blue flame, zinc and mercury with a turquoise flame, etc.

Light is a vibratory radiation—visible light represents the oscillatory band corresponding to the 49th octave of vibration. Each succeeding higher octave has twice the vibratory rate per second as its predecessor—thus, starting with two vibrations per second as the first vibratory octave, we go up the scale, 4, 8, 16, 32 vibrations per second until we come to the 49th doubling which gives the figure of 562,936,846,221,312 vibrations per second and comes very near the centre of the visible light band. It is the vibration of a certain shade of green.

The difference between the different colours of visible light lie in their vibratory rate; red, the colour with the lowest vibratory rate, having an oscillation of 436 trillion times per second, while violet, the colour with the highest vibratory rate, has an oscillation of 731 trillion times per second.

We have seen that each element (and it is from elements that all compounds and chemicals are made) has its own characteristic place in the spectrum of light. The spectrum is the sequence of colours, often spread out over a wide area by instruments for closer visual observation. The difference between one point on the visible spectrum and another point on the spectrum, is solely a difference in vibratory rate or frequence of vibrations per second. The difference between different colours is likewise solely a matter of vibratory rate. The different elements produce different colours when disintegrated, which is another way of saying that each element, metal or chemical has its own vibratory rate and the colour produced is representative of that vibratory rate. All visible colours have a vibratory rate in the range of 436 to 731 trillion oscillations per second. All elements likewise have spectral lines within that range.

In succeeding chapters we shall show the action of different colours upon the different organs and systems of the human body, followed by the reasons given by advocates of colour-therapy as to why it is preferable to apply the colour rays representative of chemicals rather than to administer the chemicals or elements directly in drug form.

Chapter 2   THE BASIC COLOURS OF METABOLISM

It is well known to biologists, doctors, etc. that the human process of metabolism is a balance between the two processes of Anabolism and Catabolism. To define these terms, anabolism is the building process by which energy is produced, tissues created and maintained, damage to cells from accident or disease is repaired, etc., while catabolism is the opposite process which is necessary to keep growth from running wild, and to eliminate waste products from the body.

In degenerative diseases, anabolism is weak and catabolism has the upper hand, which means that tissue repair is inadequate and organs therefore degenerate.

In toxic diseases, catabolism is deficient, with the result that toxic or waste substances pile up in the body and poison the entire system.

In cancer and other tumorous conditions, anabolism is excessive and catabolism low, so that the growth principle gets out of hand.

Good health can be maintained only if a proper balance is kept between the two processes of anabolism and catabolism, which together represent metabolism.

Ghadiali terms the two opposite processes as construction and destruction. In the human body, it is found that red is the colour of construction, and activates the liver. Similarly, violet is the colour of destruction, and activates the spleen.

These principles are significant in another way, for the red corpuscles of the blood are produced in the bone marrow only with the aid of certain secretions from the liver, and liver extract has long been known as an essential remedy for extremely anaemic conditions, anaemia representing simply a deficiency of red corpuscles in the blood. The new vitamin B-12, of additional aid in combating anaemia, is likewise a red substance.

It is also significant to note that there is a continuous creation of new red blood corpuscles in the living body—which means that since the number of red cells in the blood remains relatively constant when one in in good health, there must be some place where the older red cells are destroyed or taken out of circulation. Where do we find that place? In the spleen Another indication of the correctness of Ghadiali's principle that the red ray is the colour of construction and the riolet ray, which activates the spleen, is the colour of destruction or catabolism.

But there are other corroborating factors. The so-called white cells of the blood (actually violet, according to Ghadiali) are produced in the spleen. These white cells have for their function the destruction of harmful bacteria in the body, which is why the number of white cells produced rises sharply during periods of infection. So, the white cells are direct agents of the principle of catabolism, and their production is stimulated (if the spleen is deficient in functioning) by irradiating the splenic area with the violet ray, the ray which chromo-therapy terms the colour of destruction or catabolism.

It is necessary for good health that a proper balance be maintained between these two basic principles, and therefore between the action of the liver and spleen which represent polar opposites both in function and in the colour of light that activates or stimulates these two organs. Red, which stimulates liver activity, is at one end of the visible spectrum of light, while violet, which stimulates splenic activity, is at the other end of the spectrum of visible light.

The central or balancing colour of the spectrum of visible light is green. In the solar spectrum there are three colours on each side of the green. Is it not interesting that green is the colour which activates or encourages the activity of the pituitary gland? For the pituitary has long been known to be the controlling or master gland which regulates the action of all other glands in the body, and through those other glands, the action of every part of the body. It is particularly appropriate, then, that green, the central colour of the visible spectrum, is the ray which fosters *balance* in the body between the opposing actions of the liver and the spleen, between anabolism and catabolism, and that this is accomplished through the medium of the pituitary gland.

From the foregoing, it will be understood that red, green and violet are the primary colours in chromo-therapy. Contrary to popular belief, they are also the primary colours of visible light! Red, Yellow and Blue are the primary colours when working with pigments, but light rays follow different laws from those that apply to the mixing of pigments.

The first scientist to publish the theory that red, green and violet are the true primary colours, was Dr. Thomas Young of Milverton, Somerset, England, whose work on the subject was published in 1804. It was corroborated by von Helmholtz in the nineteenth century and by J. Arthur Hyatt in the twentieth century. Of the many colour theories tested by Ghadiali, it was the only one proven correct by intensive experiments. Proofs that red, green and violet are the primary colours of light, will be found in an appendix at the end of our section on chromo-therapy.

Chapter 3 ADDITIONAL PROPERTIES OF THE PRIMARY COLOURS

Red. The colour which stimulates the sensory nerves, and therefore is of benefit in deficient action of any of the five senses—sight, hearing, smell, taste, and touch.

The use of the red ray in anaemia has already been implied, through its action in stimulating liver activity.

Green, carrying the chemical potency of chlorine, a disinfectant and germicide, is the germicidal and purifying colour. Irradiation of green during a cold or any infectious disease helps the body to combat the infection.

Green is also the colour equivalent of nitrogen. Nitrogen is the largest component of proteins, which build muscle and other cell tissues in the body.

Green is found to balance not only the physical processes of the body, but also the emotional processes. When used on the head when one is emotionally disturbed, a state of calm is much more easily regained.

Violet, besides being the splenic stimulat colour, is both a motor nerve depressant and a lymphatic depressant. This makes it useful in calming or overcoming the excesses of violent insanity, in controlling excess irritability in the sane, and in overcoming excessive or morbid hunger.

Chapter 4 THE USE OF COLOURS FOR INJECTION, EXTRACTION, AND BALANCING OF VIBRATORY ENERGIES

An understanding of the functions of the three basic colours of metabolism—red, green and violet—is essential for an adequate insight into the manner in which chromo-therapy functions. The premise is that all disease, other than congenital defects or the results of injuries from accidents, represents the effect of one or more of the following causes:

(*a*) Excess of colour-energy on one side of the spectrum.
(*b*) Deficiency of colour-energy on one side of the spectrum; either a. or b. leading to an unbalance in the body between the energies from the two sides of the spectrum.
(*c*) Deficiency in the function of the balancing mechanism within the body.

With green as the central balancing point, we will term all colours on the red side of the spectrum as the hot colours, namely: red, orange, yellow, lemon and scarlet; and all colours on the violet side of the spectrum will be termed the cold colours, namely: turquoise, blue, indigo, purple, and violet. Various other terms colud be and have been used, but this terminology is simple and clear—we know that below the red lies the infra-red or heat range, and that red is a stimulating or hot colour.

We now come to a very basic or fundamental principle in the use of chromo-therapy. In an ailment indicative of an excess of hot-coloured energy in the body, beneficial results are obtained by using one of the cold colours in treatment. For example, inflammations and ailments with fever represent an excess of heat in the body, and are noticeably relieved by the administration of the blue light ray, which is from the cold side of the spectrum.

Similarly, in an ailment indicative of an excess of cold-coloured energy in the body, the administration of one of the hot colours is of aid.

The use of the two aspects of this principle is represented clearly in the treatment of burns. Burns are of two types—those caused by heat, which comes from vibrations slower than the frequencies of visible light (below the red end of the spectrum) and those caused by the use of radium, X-rays, frostbite and refrigeration, which come from vibrations faster than the frequencies of visible light (above the violet end of the spectrum.)

The vibratory rate, or frequency of oscillation, of the energy that *produced* the burn, gives the key to the colour-energy required for treatment. Burns caused by heat, or excess of hot energy, are relieved and more rapid healing is promoted, through the administration of the blue light ray, an energy from the cold side of the spectrum of visible light. On the other hand, burns caused by radium or X-rays result from an excess of cold energy, and are relieved and more rapid healing is promoted, through the administration of the red ray, which supplies the *deficiency* of hot energy that has occurred in the body because of the burn.

Burns from sources of cold, such as frostbite and refrigeration, act through rapid *depletion* of the hot energies in the body, and are likewise relieved by the use of the red ray, which aids in making up the deficiency.

The above illustrates the use of colour energies in healing *by injection* of the vibratory energies in which one has become deficient. Now let us turn to healing *by extraction* of excess energies.

In all fevers there is a predominance of hydrogen and carbon in the body. Both of these chemical elements have colour energies on the hot side of the spectrum. The spectroscopic colour of hydrogen is red, and of carbon is yellow. To reduce the fever, the excess hydrogen and carbon must both be burnt out. How? By combustion—and the one element that has *affinity* for both is oxygen. The spectroscopic colour of oxygen is blue. The use of the blue light ray has a very soothing and healing effect in fever conditions, and in fact in all inflammatory conditions. Any inflammatory condition is indicative of an excess of hydrogen in the body.

Oxygen unites with hydrogen to form water, a harmless neutral compound. Oxygen unites with carbon to form carbon-dioxide, which is eliminated from the body through the lungs during the periods when we exhale.

Thus, the administration of the blue (oxygen) energy has the effect of *extracting* the excessive hydrogen and carbon (red and yellow) energies, through combination, neutralization, and then elimination from the system.

In all cases where balance of colour energies is obtained through extraction of excess energy of a particular colour, it is the "affinity" principle which makes this possible, i.e., there are affinities between the energies of certain colours having opposite attributes, just as there are affinities between certain elements which result in their combination into neutral compounds.

Affinity in chemistry refers to attraction. In chromo-therapy, affinity waves have opposite attributes or qualities, hence they seek one another to combine or neutralize. Just as hydrogen and oxygen have an attraction for one another and rush towards combination (on the application of a spark) to produce water, *so do the attuned colour waves absorb affinity waves wherever they be found in the body, and convert them into neutrality.*

The affinity pairs; that is, the waves having opposite attributes are:

Red and blue
Orange and indigo
Yellow and violet
Lemon and turquoise
Scarlet and purple.

It will be noticed that the left member of each of the above pairs is from the hot side of the spectrum, while the right member is from the cold side of the spectrum.

In any illness, there is an unbalance in the body of the energies from the cold and hot sides of the light spectrum—for in a body where these energies are in balance, a condition of health prevails. The nature of the illness shows which energies are deficient or excessive, thus pointing the way to the selection of the colour energy or energies that will work in the direction of correcting the deficiency or excess which is causing the illness. However, it is also advisable to strengthen the balancing powers of the body itself, which is accomplished through administration of the green colour as previously mentioned. Another balancing colour is magenta; the uses and function of this colour will be described later.

Chapter 5   THE SECONDARY COLOURS

The three secondary colours are each half-way between a pair of primary colours. Half-way between the vibratory rate of red and green we find the vibration which produces the colour *yellow*. In fact, two light beams, one red and the other green, if allowed to focus on the same spot, will merge into yellow.

Likewise, half-way between the vibratory rates of green and violet we find the vibratory rate which produces the colour *blue*. Two light beams, one green and one violet, if focussed on the same spot, merge into blue.

Similarly, half way between the vibratory rates of red and violet, we find the vibration which produces the colour *magenta*. If a light beam is passed through two slides, one red-coloured and the other violet-coloured, the result will be magenta. Magenta has the same numerical vibratory rate as green; the difference between green and magenta will be described in a later chapter.

Chapter 6   PROPERTIES OF YELLOW

Yellow is the colour that activates the motor nerves and is therefore the energy generator for muscles. Any disturbance in the supply of yellow energy to any part of the body, can cause disturbance of function in that area, including partial or complete paralysis. Regardless of the type or apparent cause of paralysis, the basic cause is the deficiency of sensory and/or motor energy. The remedy is on the hot side of the spectrum, and usually yellow. Some surprising cures of supposedly hopeless paralysis cases have been accomplished through daily irradiations with yellow light.

It is interesting to note that remedies which have been used by physicians to try to help paralytic conditions, are heavily predominant in hydrogen and carbon elements—which represent the red and yellow energies (sensory and motor stimulants) respectively. For example, the formula of strychnine is H22 C21 N2 O2, meaning 22 atoms of hydrogen, 21 atoms of carbon, 2 atoms of nitrogen, and 2 atoms of oxygen to each molecule. Notice that hydrogen and carbon predominate. Similarly, the formula of brucin is H26 C23 N2 O4. Ghadiali makes the observation that both of these compounds irritate the spinal cord and can cause convulsions. Also that although they are powerful stimulants, they have very little of the reparative potency, which resides in nitrogen, the green ray. Thus as medicine, they have drawbacks.

Yellow, being a mixture of red and green vibratory rates, and in fact directly producible by a combination of red and green beams, has half the stimulating potency of red and half of the reparative potency of the green or nitrogen vibration, thus it tends to both stimulate function and repair damaged cells.

Yellow, due to its stimulating action upon all muscles, is of aid in energizing the alimentary tract and in improving digestion when latter is sluggish. The yellow light directed at the intestinal area for short periods is a digestant. For longer periods there tends to be a cathartic action. It also stimulates the flow of bile, and has an anthelmintic action (antagonistic to parasites or worms).

Chapter 7   PROPERTIES OF BLUE

Blue is one of the most useful of the colour rays, due to its action previously described in combating inflammation and fever, and in the aid for healing burns caused by an excess of heat energy. Blue also helps to relieve pain, allay itching, and, through carrying the potency or vibratory correspondence of oxygen, it acts as a builder of vitality. It has a calming and soothing influence, and while not as sedative in effect as indigo, it still can have a marked effect in aiding some individuals to overcome insomnia.

Chapter 8   THE TERTIARY COLOURS

These are orange, lemon, turquoise, indigo, scarlet, and purple. With the primary and secondary colours already listed, the tertiary colours complete the group of twelve colours used in the Ghadiali system of chromo-therapy.

Orange, is half-way between the vibratory rates of red and yellow and is produced through the combined use of red and yellow slides.

Lemon, is half way between the vibratory rates of yellow and green, and is produced through the combined use of yellow and green slides.

Turquoise, is half way between the vibratory rates of green and blue, and is produced through the combined use of green and blue slides.

Indigo, is half way between the vibratory rates of blue and violet, and is produced through the combined use of blue and violet slides.

Scarlet and purple will be discussed when the section on magenta is reached.

Chapter 9   PROPERTIES OF ORANGE

Orange is the colour which stimulates or energizes the thyroid gland. Due to the action of the thyroid in stimulating the function of respiration, orange is a respiratory stimulant. It is also a depressant of para-thyroid action. The para-thyroid glands are opposite in function to the thyroid, and in good health there is a proper balance between the function of the single, large thyroid and the four, small para-thyroids. The relationship between the actions of thyroid and para-thyroid respectively, controls breathing. The orange vibratory impulse from the thyroid expands the lungs; the indigo vibratory impulse of para-thyroids contracts the lungs.

Thyroid with its orange activity, accentuates the hydro-carbon side of human chemistry in metabolism through oxidation. Remember that hydrogen is the red potency and carbon the yellow potency; combining the two we have orange. It is therefore quite appropriate, and significant to the accuracy of the Ghadiali chromo-therapy method, that the gland which stimulates metabolistic oxidation in the body is activated with the orange light ray.

For thyroid-deficient action, the use of the orange ray is indicated. For over-active thyroids, the indigo ray is used, since indigo is both a thyroid depressant and activator of the para-thyroid glands which work in opposition to the thyroid.

Orange also has an anti-spasmodic effect, so is useful in the relief of cramps or muscle spasms of all kinds. The writer has seen surprising demonstrations of this, where it took the place of heavy sedatives. As orange is the spectroscopic colour of calcium (calcium in the process of disintegration produces lines in the orange portion of the light spectrum) the use of orange aids the calcium metabolism of the body, and strengthens the lungs.

Another use of orange light is to tone up the stomach when the latter is not functioning well. If there is abnormal putrefaction in the stomach, orange can have an emetic effect, if the stomach is first filled with water. If there is no abnormal putrefaction in the stomach, then orange does not have an emetic effect.

Orange also stimulates the milk-producing action of the breasts after child-birth.

The use of orange will be further discussed in the chapter on tuberculosis.

Chapter 10   PROPERTIES OF LEMON

Lemon-coloured light acts as a cerebral stimulant. It is known that phosphorus and sulphur stimulate the brain, also that gold in the male and silver in the female have a basically stimulating effect for both the brain and sex systems. It is interesting to find, therefore, that the spectroscopic colour for all of these elements (phosphorus, sulphur, gold, and silver) is lemon. In other words, when administering the lemon-coloured ray, one is giving the same vibratory rate of frequency which corresponds to the various chemical elements which are known to stimulate brain function.

Lemon also activates the Thymus gland. Ghadiali states that the Thymus gland, located in the upper part of the chest, contains uranium in its composition at birth. We know that the thymus is a regulator of physical growth and maturity. Ghadiali states that these processes are due to the slow disintegration of uranium, which produces chemical elements necessary to the growth of the body, and that when the store of uranium in the thymus is used up, no further appreciable physical growth in the body can occur.

Lemon is the colour-ray advocated for use in cretinism, dwarfism, etc., in view of its effect in stimulating the thymus gland.

The lemon colour, like the lemon fruit, is considered to have an antacid effect in the human body.

However, perhaps the most important use of the lemon-coloured ray is as a chronic alterative (remedy for chronic disease). Lemon has the radiant energy of gold and iodine, both of them having a long history of use as alteratives. Lemon, being half-green, has the effect of a cleanser of the system, and being half-yellow, also has the effect of motor stimulant to throw off morbid debris (waste products, toxic accumulations, etc.). Ghadiali states "It has no equal for rejuvenating the organism in cases worn by battle against disease." He further explains that in persistent disorders—those that have continued for a long period of time, the energy required to destroy pathological processes has been spent, and needs rejuvenating or reviving, which is a function of the lemon ray.

The lemon light can be used as a laxative, since it has half the potency of yellow, a cathartic colour.

In cases where there is coughing, there is a need to eliminate phlegm. This requires motor stimulant (yellow) and cleanser (green). Lemon gives both, when irradiated over the affected areas.

Lemon also helps to build and strengthen bones, through its correspondence with the phosphorous vibration.

Chapter 11   PROPERTIES OF TURQUOISE

Turquoise, produced by combined use of green and blue slides, has advantages over the use of blue in recent disorders as an "acute alterative", since it combines the cleansing effect of green with the soothing effect of blue, though blue is the paramount colour to use during the period when fever is excessive. After the fever has lessened, turquoise is the colour-ray advocated for use.

Turquoise has the opposite effect from the lemon-ray. Turquoise is acid and tonic in its action. The acid attribute of turquoise is drawn from its proximity to the blue, oxygen wave. Oxygen is the essential element in all acids except five. Those five instead, are made with hydrogen; hydrofluoric, hydrochloric, hydrobromic, hydriodic, and hydrocyanic acids. Of these five, the most corrosive is hydro-fluoric acid, of which the major element is fluorine. Spectroscopically this is in the turquoise group.

Turquoise is the prime skin-building colour. After the pain from burns has been conquered by the use of the blue light, the use of turquoise hastens the formation of new skin.

Another function of turquoise is as a cerebral depressant, and it is therefore used to quiet mental over-action.

Chapter 12   PROPERTIES OF INDIGO

This colour ray is a para-thyroid stimulant, and a thyroid and respiratory depressant as previously outlined. It also has marked sedative and pain-relieving qualities.

Another very important use of indigo is for its haemostatic ability, i.e. to help reduce or stop excessive bleeding. While it cannot stop blood flow if a major blood vessel has been severed, it has been demonstrated to have a very beneficial effect on blood seepage where healing of more superficial wounds is inadequate, and, when properly used, in reducing or stopping the loss of blood in cases of excessive menstrual flow.

Another function of indigo is to stimulate the formation of phagocytes (one of the types of white blood cells) in the spleen.

Having made a rapid survey of the known effects of the different colours from red to violet in the visible light spectrum, the reader by now would probably like to obtain an insight into the mechanism by which the body obtains from coloured rays the various benefits that have been outlined. So we turn to that division of the subject, before detailing the use of colours in certain major and common ailments.

Chapter 13   HOW COLOUR-RAYS PRODUCE PHYSICAL EFFECTS

In one of the introductory chapters to *Encyclopaedia of Spectro-Chrome-Metry*", Ghadiali wrote in part:—

> "We are more than physical machines, since we can rebuild worn parts within ourselves. We are a marvellous mechanism of self-repairing living factors.
>
> The live man and the dead man have a certain difference in their composition and behaviour.
>
> There is something within man which is more than sheer chemical body; and Spectro-Chrome is the system of healing which . . . introduces the remedy for the disorder into those Higher Vehicles through which man functions as a live entity.
>
> To understand Spectro-Chrome one must first learn about the higher vehicles of man."

There have always been a few individuals down through the ages who had the faculty of seeing the human auras—luminous bodies surrounding and interpenetrating the physical body, and composed of very tenuous substance vibrating at higher frequency rates than were visible to the normal eye. These auras were of different colours in different people, and various types of disease or ill health were accompanied by characteristic discolorations in appropriate portions of the auras of the individuals involved.

During the decade of 1910 to 1920, the British scientist Walter J. Kilner developed methods for making the auras visible to almost everyone, through the use of transparent screens containing dyacyanin. This chemical has the property of radiating an emanation which has a tendency to temporarily alter the characteristics of human sight so that higher-than-normal frequencies can be observed. Kilner's observations and methods are described in detail in his book *The Human Atmosphere* published in London in 1920 and re-printed in 1926. His work established the presence of auras as a scientific fact. Others who have written books on the aura include Bagnall, and Dr. George Starr White.

In recent times, the substance of dyacyanin has become difficult and at times impossible to procure. A number of groups of individuals interested in auric observation are developing special lenses and other devices so that the auras can again be made visible to those who would not otherwise be able to observe them.

With the special apparatus, at least two auras can be discerned, which are considered to correspond with the etheric and astral bodies of man, much discussed in occult literature.

Many of the principles embraced by metaphysical and occult study have since become proven facts through scientific research. For example, the existence of the planet Pluto was known to occultists for a long period before its presence was located by astronomers. The pioneering research at Duke University has established the factual existence of various faculties hitherto considered in the range of the occult—they are now termed "extra-sensory perception".

The occult view is that the higher bodies of man, or auras, receive from the atmosphere the white-light energies, and these energies are then processed and distributed to various parts of the physical body. Colour-rays are deemed to act upon the auric bodies of man, through which they then influence the physical functioning.

In order to obtain a better understanding of how chromo-therapy functions, it is necessary to examine the nature of matter, and also the nature of light.

Chapter 14   THE NATURE OF MATTER

The molecule is the unit of matter. There are many thousands of types of matter, but these are all composed of some ninety-eight elements. The unit of an element is the atom. Different combinations of atoms combine into molecules of the various types of matter that are known.

The above is the conventional view. Ghadiali, on the other hand, stated that elements themselves were compounds, and that this statement is proven by the fact that none of the known elements have a single or pure spectrum. He further stated that there is a *constant tendency of matter to disintegrate, and that all matter radiates* as a result of this disintegration process. As one of the proofs, he cites the fact that each metal has an odour distinct from others. Odour is stimulation of the olfactory tract by particles in the air. If metal was not disintegrating, it would be unable to issue an odour.

L. E. Eeman, whose work will be outlined in later pages of this book, proved by experiments that substances such as drugs, minerals, bacterial cultures, etc., have radiations which can be conducted upon wires so as to affect the human organism. Proof of the validity of Ghadiali's concepts is accumulating.

Long before physicists had discovered how to split the atom, Ghadiali, writing in 1933, forecasted this event. He further stated that the then current view that atoms were composed of only a few types of electrical particles, was incomplete, and that future discoveries would show the atom has many components. Since 1933, several additional components of the atom have been discovered.

The Nature of Light

Sir William Crookes demonstrated fluorescence and phosphorescence of minerals and other matter by placing each in a highly evacuated glass bulb and focusing high tension electrical discharges upon it. This speeds up the natural processes of disintegration and accelerates the radiation of matter. Each material becomes radiant under such bombardment. *Light is a function of the disintegration of matter* The colour of light, and therefore the vibratory range represented in the light, is a function of the type of matter that is disintegrated. For example, a Cooper-Hewitt Mercury vapour lamp gives colours from green to violet, with no red. In contrast, a neon tube gives the red section of the spectrum. (When other colours than red are emitted by tubes that are termed "neon", it is because other gases have been added.)

Sunlight, carbon-arc lamps, and incandescent bulbs give the complete visible spectrum range, and therefore represent the entire vibratory octave of visible light including all of the constructive chemical potencies, as outlined on page 49.

The sun actually gives no light, but sends *Energies* arising from the processes of combustion of matter—and these energies are convertible into light when they pass through the impediments in our atmosphere. Colour is nothing but a divisional part of light; a narrower portion of the vibratory range of the forty-ninth octave of vibration.

Michael Faraday proved in 1845 by polarized-light experiments, that sound, heat, light and magnetism are all the same energy, differing only in frequency of vibration and in the medium of conduction.

All known energies are composed of oscillatory frequency in different media of transportation, and all life (as contrasted to the matter in which the life manifests) is composed of energies. Therefore it should now be easier to understand how the administration of beams of light can reinforce the life energies of the human body.

Another proof that light contains energy, is found in Sir William Crookes' experiments with radiometers. The radiometer is a glass globe with very high vacuum. Inside are pivots at top and bottom. A "weather vane" with four vanes is mounted on a vertical shaft that turns in the pivots. When light is thrown on the globe the vanes rotate as if struck with a physical force. Motion cannot be produced without energy. In this case the energy is supplied by the light.

To recapitulate:

The vibratory essences of a drug, chemical or element are represented in the colours of light thrown off when that drug, element or chemical is disintegrated.

Corresponding therapeutic effects are obtainable from using the colour of light that correlates with the elements in which the body is deficient, or which have an "affinity reaction" with the elements with which the body is over-supplied.

The different colours of light represent different energies required by the human body for its proper functioning.

The light rays are absorbed by the auric bodies of man, through which the remedial effects on the physical body are obtained.

Colour rays are much more than mere substitutes for drugs, in the opinion of the advocates of chromo-therapy, as will be seen in the following chapter.

Chapter 15   ADVANTAGES OF CHROMO-THERAPY OVER THE ADMINISTRATION OF ELEMENTS OR COMPOUNDS IN MATERIAL FORM

When we analyze the chemistry of man, we find the following average composition:

By elements:		By components:	
Oxygen	72. %	Water	84 lb.
Carbon	13.4	Carbon	44 lb.
Hydrogen	9.1	Ammonia	8 lb.
Calcium	1.3	Calc. Carbonate	53 oz.
Phosphorus	1.25	Phosphorus	28 oz.

and small fractions of sulphur, sodium, chlorine, fluorine, potassium, iron, magnesium, silicon, copper, lead and aluminium.

Ghadiali states: "Thousands of drugs are used in medical practice. Is it wise to dump so many into the human body when they were not included in the natural composition of the body?

"For example, there is no perceptible quantity of mercury in the human composition, yet this poison is administered in large quantities by doctors for syphilis and other ailments.

"Percentage of iron in the whole body is only .01—about 77 grains. Yet doctors give far greater doses in a month's time. Overdosing naturally produces undesirable effects.

"Medicine ignores wholesale, the fundamental chemistry of the human body—pouring into it many drugs containing elements not found in the body, or in quantities far in excess of their natural proportion in the body."

"NO PART, NOT BUILT FOR FUNCTIONING IN A MACHINE, CAN BE SHACKLED INTO IT WITHOUT UPSETTING ITS RHYTHM. CHEMICALS ARE LIVE POTENCIES; THEIR ATOMS HAVE ATTRACTIONS AND REPULSIONS, AND TO ENDEAVOUR TO INTRODUCE HAPHAZARD INORGANIC METALS INTO AN ORGANIC MACHINE, IS LIKE FEEDING A BABY WITH STEEL TACKS TO MAKE IT STRONG."

Another point made by Ghadiali in this respect, is that deviation of an element in the body above or below its normal percentage is a prime cause of illness, and physicians often unwittingly increase the imbalance or reverse it to the opposite side with their remedies—hence in trying to cure an ailment or a set of symptoms, they often induce other ailments. For example, salicylate given for rheumatism, can affect the heart adversely; then if digitalis is given for the heart, kidney function is likely to be impaired, and the medicine given to help that is likely to cause harm elsewhere in the body. Few doctors, according to this view actually comprehend the true effects of the drugs they administer. The so-called "cure" is mostly a shifting of disorder from one organ to another.

Other objections mentioned, were that drugs can be unreliable; the same drug can have different effects on different people—also, that too many toxic drugs are used, damaging instead of rebuilding the body.

In contrast, chromo-therapy leaves no harmful residues which the body has to struggle to try to eliminate. Chromo-therapy reaches directly to the fundamentals of illness, by re-establishing a balance among the vibratory energies in the body, and it is these vibratory energies which activate all the different organs, glands and systems of the body. Chromo-therapy is much more than simply an indirect means of affecting the physical body. The auric bodies are integral parts of man's constitution and play the major part in supplying the physical body with the energy needed to sustain life. Many of the ailments to which man is subject, have their origin and primary stronghold in the auric bodies, as outlined in detail in the book *Some Unrecognized Factors in Medicine*, published in London, 1939. In order to treat the auric body successfully, it is necessary to use modalities having low power and high rates of vibration. The ordinary modalities based on chemical reactions or the more crude or destructive vibrations, are ineffectual if not actually injurious to the auric bodies. In applying chromo-therapy one is using the type of remedy which most closely matches the characteristics or constituent materials of the all-important auric bodies. When a poisonous drug or element such as Mercury is administered, whatever constructive effect is obtained is due to the colour potency of the substance—in this example, turquoise. By giving the colour-ray to the body instead of the drug, it is the premise of chromo-therapy that the constructive effect can be obtained without any accompanying destructive effect, for there are no poisonous colour rays from the visible light spectrum. Suppose that a mistake is made and an incorrect colour is used or too long an exposure is given—the worst that can happen would be a temporary accentuation of functional disorder; unlike the results of overdose of infra-red or ultra-violet, no tissue destruction can take place.

It is curious to note that the medical profession is quite willing to use the vibratory portions of the light spectrum just above and just below the visible light band—namely the ultra-violet and infra-red portions, yet is slow to acknowledge the constructive healing qualities of the visible light spectrum. Both infra-red and ultra-violet rays cause burns and other tissue damage if used to any sizable extent, in contrast to the results from rays from the visible portion of the light spectrum. The few medical physicians who have studied chromo-therapy thoroughly and have applied it adequately, are well pleased with its results.

We shall now take up the study of three colours that are not found in the solar spectrum or the rainbow, but which are of profound significance with regard to the functioning of the human body.

Chapter 16   THE SECRET OF MAGENTA

It will be seen that the three primary colours of light—Red, Green and Violet—constitute a triangle in an ovoid or ellipse. This triangle is not arbitrary; it represents the vibratory relationships involved. Each secondary colour is formed by combining the two primary colours, one of higher vibratory rate and the other of lower vibratory rate than the resulting secondary colour. The vibratory rate of the secondary colour is the exact mathematical average of the vibratory rates of the two constituent primary colours.

Just as yellow is the secondary colour between red and green, and blue is the secondary colour between green and violet, so is magenta the secondary colour between red and violet, and is produced if a beam of light is conducted through two slides, one red and one violet. (In practice, it is preferable to use a single, magenta-coloured slide, as the violet slide has such low luminosity that the *quantity* of light passing through is drastically reduced.)

The peculiar aspect of magenta is that it has the same vibratory or oscillatory rate as green, yet the visible colour is different from green, and the effects upon the body are likewise different.

How can two colours be different and at the same time have the same vibratory rate? The answer is that the vibratory rate is not the only factor of colour, just as the numbers of atoms are not the only factors in the construction of molecules. Kerosene, and attar of roses both have the same chemical formula—$C_4 B_4$. Coal and diamonds have the same formula of constituent atoms, but what a difference! In matter, the difference lies in the way in which the atoms are assembled together. The same atoms assembled in different sequences produce different results. With light, the difference lies in the direction of rotation with respect to the plane of polarized light. Magenta has the opposite direction of rotation from green.

Why are we so interested in magenta? It is because this is the colour which energizes the adrenal glands, the heart action, and the sex mechanism. Ghadiali states "Green, through the head, forms the North pole of the human body; the magenta, through the genitals and allied sex mechanism, forms the South pole of the human body. Thus under the laws regulating electro-magnetism, etc., both form in actual functioning a linked apparatus, which works as one single factor. . . . No head, no sex; no sex, no head; this refers not to sex congress, but to the fact that the endocrines, or internal secretion products from the glandular structure, operate in unison and as one mechanism, and not separately as many would believe. Thus, the functions of both green and magenta go hand in hand—same as the poles of the magnet which can never be isolated, being dual in manifestation".

Magenta reinforces the heart action both directly and indirectly; directly, if used over the heart area, through its correspondence with the vibratory rates of lithium, potassium, strontium, and manganese, all of value in energizing the heart; indirectly, through the function of stimulating the supra-renal or adrenal glands, when used over the area of those glands, which in turn have a stimulating effect upon the heart. Magenta's vibration not only tends to strengthen the heart muscle but also tends to stabilize the heart rhythm. There is a diuretic effect, and attributes similar to the constructive aspects of the effects of digitalis and strophanthus.

The sex potency of humans depends partly on the amount of blood circulation through the genital area, hence upon the vigour of the heart in driving the circulation, and sex potency also depends upon auric strength. Both of these derive mainly from magenta, which is the predominant colour of the normal human aura, according to Ghadiali. He further states that no individual of giant intellect was known to be of neuter gender. There is no stigma attached to sex virility—only to its abuse. Sex virility can be used for great thoughts, great words, and great deeds. It is that which gives "pep", as well as being an important factor in what is termed "personality".

In the discussion of magenta, the point is made that it is the power of this vibratory rate which draws people together in love—meaning the genuine human affection. This same oscillatory frequency, when unbalanced, produces hysteria in women. Prevalence of female disorders in the present age is considered due to the unbalancing of magenta, and even insanity in some cases, is traced to this cause.

Because of the part played by the head and the spinal column (representing the North-South pole axis of the magnetic system of the body) in any disorder, it will be found beneficial to start colour therapy treatments with the green ray given systemically and the magenta ray given locally over the genital area.

To sum up, the three areas of the body on which the magenta ray has useful effects are the heart, genitals, and supra-renals (located on top of the kidneys).

The effects of magenta, scarlet and purple on heart ailments will be discussed in the chapter on heart disorders. Scarlet and purple have other uses, listed below.

Chapter 17   PROPERTIES OF SCARLET AND PURPLE

We shall discuss these two colours together, since they have precisely opposite effects. Scarlet has a vasoconstricting effect, and tends to raise blood pressure, while purple has a vasodilating effect, and tends to lower blood pressure. Scarlet stimulates kidney activity and can be used for that purpose when kidney function is underactive; purple depresses kidney function, and can be used for that purpose when kidneys are overactive. Scarlet stimulates the sex mechanism, helps overcome impotence and frigidity when used on the genital area, and builds sex potency in those whose sex impulse is subnormal. Purple depresses the sex functioning, reduces sex sensitiveness, and therefore is useful in conserving the sex force and building sex potency in those whose sex functioning is over-active. Scarlet is of aid in cases of scanty menstruation; purple has the opposite effect, so is used where menstruation is too profuse. In more severe cases of excessive flow, indigo is indicated for its property of reducing bleeding.

Purple has an analgesic, anti-pyretic, narcotic and hypnotic effect, and for those purposes can be used with a rather long exposure. Purple is also said to have an anti-malarial effect.

Chapter 18 EXAMPLES OF THE USE OF COLOUR RAYS IN TREATMENT OF SOME OF THE COMMON AILMENTS

In presenting these examples we are simply reporting the methods advised by chromo-therapy advocates, as found helpful by doctors who have used this system in practice. We are not attempting to "prescribe" for anyone, nor do we either claim or disclaim that colour rays will "cure" any ailment. The writer has had personal experience testifying to the effectiveness of chromo-therapy in certain functional disorders, and is acquainted with doctors who have found this type of therapy successful in the treatment of burns which had not yielded to any other method of treatment, and in certain organic conditions. It is the viewpoint that chromo-therapy is a subject which deserves further research and investigation which it is hoped will be stimulated by this current publication of the principles of chromo-therapy as developed to date.

Colds

The colour-ray to use is green on the head, and then on any other area that manifests symptoms of the cold. Green is used for its bactericidal, cleansing effect; also for its balancing effect through stimulation of the pituitary gland. Immediately following the application of the green light, blue can be used on any area in which inflammation is present—head, throat, chest, etc. The use of blue in combating inflammation has been discussed on a previous page. The same sequence of colours has been found of benefit in sinus attacks.

Tuberculosis

Ghadiali in his encyclopaedia writes of the success of chromo-therapy in arresting tuberculosis in many cases that had not yielded to other treatment. In his analysis of the process of tuberculosis, he points out that lungs are composed of spongy material. If this material breaks down, circulating and surrounding fluids escape from their proper channels. To stop leakage in the lungs, calcium is needed—just as masons repair breaks with cement (lime). Copper also plays a part in the maintenance of lung tissue.

Turquoise (acute alterative) is specified for use during the period when fever is present. If the fever increases, switch to blue. Rest and diet are also important and should receive adequate attention. After the fever has declined almost to normal, the main remedy in chromo-therapy lies in the use of the orange ray—carrying the potency of both calcium and copper, the two elements most needed in repairing lung tissue. The orange ray to be irradiated on both front and back of the lung area.

In discussing the cause of tuberculosis, Ghadiali reviews the germ theory and states that comparative microscopic examination of animals at death and again at an interval following death, discloses many forms of small living organisms at the latter examination that were not present at the initial examination. He states that microbes are the results of various processes of decomposition, decay, and destruction. They do not *cause* imbalance in the body—the imbalance *produces* them. This viewpoint has since been upheld by the new methods of microscopic investigation developed by R. R. Rife and by Dr. Wilhelm Reich, M.D., both of which will be reviewed in later sections of this book. Rife, with his new types of microscopes has been able to show that harmless bacteria always present in the body, can change into deadly bacteria if there is a slight change in the chemical content of the environment, and Reich showed that degeneration of body tissues resulted in the production of harmful micro-organisms.

Ghadiali's views, and the scientific studies made by R. R. Rife and Dr. Wilhelm Reich all point to the validity of the Naturopathic view, which is interested in preserving the health and vitality of man as a whole organism, and which considers that if the body vitality is maintained through nutrition and other measures, then infections and other diseased processes cannot occur. In treatment, Naturopathy involves the re-activation and re-establishment of normal functioning of the body as a whole, in contrast to medical methods which have been so concerned with fighting symptoms, as for example, finding drugs or serums to combat specific micro-organisms.

Burns

The technique for treating burns with chromo-therapy calls for dressing the burns with tissue-paper coated with coconut oil thus allowing the colour rays to play upon the surface of the burn. Blue is used to relieve the pain, and can be administered over the affected area for long periods of time when used for this purpose. When the pain is gone, turquoise is then used in a regular sequence of treatments to rebuild the skin. Chromo-therapy has accomplished outstanding results in healing burns that covered a large area of the body, where patients' condition was critical.

Skin Disorders

Skin disorders are considered to be internal in cause. They fall into two groups—the moist weeping disorders, which show a preponderance of energies from the hot side of the spectrum, and the dry scale disorders, which show a preponderance of energies from the cold side of the spectrum. The latter type need first to be converted to the moist, weeping type before they are likely to heal. For all skin disorder, it is recommended that treatments start with green systemic. For the moist weeping disorders, turquoise is the colour to use next—if the case is obstinate, change later to blue or indigo. For the dry scaly disorders, follow the green with lemon; after a period of use, if moist weeping has not appeared, switch to yellow or orange. When moist weeping appears, treat with turquoise followed by blue or indigo if necessary as mentioned above.

It is important to avoid the use of soap in skin disorders. A salve of zinc oxide in vaseline was mentioned for supplemental use to relieve cracking and itching of the skin, with the comment that zinc emanates the turquoise radiation and treats between the color-therapy treatments. Color-therapy cannot be expected to remove skin growths such as warts; nothing that is pedunculated and detached from the vascular blood supply will be affected.

Syphilis

Ghadiali's use of colour-therapy for this disorder involved the irradiation with green systemic daily for several weeks front and back, then lemon systemic for several weeks, front and back. . . . He refers to inaccuracies of the Wasserman test. This writer has seen various references in medical literature detailing many conditions where the test is known to be inaccurate, as for example after a surgical operation under general anaesthetic; after a severe infection of most any kind, and after great emotional shock or fatigue.

Gonorrhea

Ghadiali's published remedy for this ailment includes first using green systemic for cleansing; during the time when there is burning urine, turquoise or blue can be used on the affected area, then shift to lemon for persistent disorders. He states that in a few weeks the disease is eradicated from the system.

Diabetes

In his analysis of this ailment, Ghadiali points out that excess sugar in *urine* merely indicates glycosuria and does not necessarily indicate diabetes, which is the ailment in which there is excess sugar in the *blood*. He wrote at length regarding the need of the body for carbohydrates, and the harm done through carbohydrate starvation. The lung action depends on the functioning of the thyroid gland. The colour wave which activates the thyroid is orange, which is representative of carbohydrates (the mixture of hydrogen—red, and carbon—yellow). Deprivation of carbohydrates drains thyroid energy, lowers lung activity—and blood circulation, metabolism and nutrition all suffer.

Of the different types of sugar, milk sugar is termed galactose, grape sugar is termed dextrose, and fruit sugar is termed levulose; the empirical formula for all is $H12$ $C6$ $O6$, but there is a vital difference between the character of dextrose and levulose, in relation to polarized light. Galactose and dextrose are dextro-rotatory; that is, they twist the plane of polarized light to the right. Levulose is levo-rotatory—it twists the plane of polarized light to the left. Ghadiali states that sugar going without assimilation is the dextrose, and that levulose is consumed within the tissues. He advocates the inclusion of levulose in the diabetic diet to avoid carbohydrate starvation. He continues, that the basic cause of diabetes is a sluggish lymphatic system—the lymph plays a large part in assimilation and nutrition. Being a chronic disorder, the colour ray indicated is lemon systemic, followed by yellow systemic as lymphatic activator.

Boils, Abscesses, Carbuncles

Lemon systemic is advised, and irradiate affected area with orange local. When suppuration or throbbing is felt, irradiate with yellow until the area bursts open. Remove the core. Then, use green local until pus drains and a clean red cavity is left. Use turquoise local a few times, and finally seal with indigo local.

Mastoiditis

By using green systemic and turquoise local, it is stated that the condition can often be cleared up without surgery.

Heart Disorders

Magenta is specified for angina pectoris, and hypertrophy. Scarlet is indicated for stenosis, incompetence of heart valve, cardiac dilation, and bradycardia (slow pulse). Purple is indicated for excessive palpitation; being on the cold side of the spectrum, it is the colour for all heart ailments characterized by inflammation—such as myocarditis, endocarditis, pericarditis, also for aortic aneurism.

Chapter 19   TECHNIQUE OF APPLICATION

To obtain the best results from chromo-therapy, a specially constructed lamp is required. The essential features are a housing for the globe or bulb, constructed so that all the light is reflected to the front, and can emerge from the lamp only through the opening covered by slide in use. In front of the bulb should be provision for inserting and changing the coloured glass slides. The housing needs to be ventilated so that at a distance of two feet in front of the lamp, no appreciable quantity of heat is felt—the aim being to obtain the light ray without extra heat. A 400-watt globe will give better results than smaller globes. If wattage above 400 is used, the housing will need to have forced ventilation from an electric fan, and to filter out the heat it will be necessary to place between the globe and the coloured slides, a transparent chamber filled with water. The lamp must be constructed so that no light reaches the patient without first going through the coloured glass slide. The housing should be mounted on a stand with provision for tilting at various angles, also for raising and lowering.

The slides need to be of clear glass with the correct shades of colouring. Rippled glass is not nearly as suitable. See Chapter 22 for factors involved in the production of the slides.

For a low-priced set-up, some degree of results can be obtained from a photographer's 150-watt spot-lamp. These often have clips in front, into which the glass slides can be inserted, but it has been found that the slides are likely to break from heat unless longer clips are installed, which will allow at least an inch of space between the front of the lens and the glass slides. For use in a photographer's lamp, the edges of the slides should be bound, to lessen chance of breakage. The lamp should be positioned so that the slides are located two feet from the area of the body to be irradiated. Treatment should be administered in a room where no other light is admitted—if during the daytime, shades should be drawn, and light from windows blocked as thoroughly as possible.

Exposures range from three to ten minutes for colours from the hot side of the spectrum (except in very acute conditions, where a longer exposure can be used) and fifteen to thirty minutes for colours from the cold side of the spectrum, except for high fevers, severe pain, acute infections, and burns, in which cases the exposures can be for an hour or even longer. Ordinarily, treatments are given once or twice a day, but in severe or acute cases the treatments can be given more often.

The colour beam should always be directed at the bare skin. Portions of the skin over organs not to be included in the treatment, should be covered with cloth or otherwise shielded.

If the patient is to be treated in a sitting position, the lamp should be positioned so that it is at the same height as the area to be irradiated, and the beam should be directed squarely at the area. To irradiate an area on the patient's back, the patient can sit sidewise on a chair.

When a colour is made up by using a combination of two slides, less light is transmitted, so a longer exposure is necessary than when one slide alone is used.

The room should be warm enough so that the patient does not feel uncomfortable in having the skin bare for the treatment.

Slides needed are: red, yellow, green, blue, violet. It is preferable also to have a magenta slide, and if an orange slide is available it permits shorter exposures, since a greater quantity of light passes through one slide than through two.

If an orange slide is not available, orange is obtained by combined use of red and yellow slides. Lemon is produced by combined use of yellow and green slides. Turquoise is produced by combined use of green and blue slides. Indigo is produced by combined use of blue and violet slides. Magenta can be produced by combined use of red and violet slides—however, the violet slide has such low luminosity that very little light is transmitted through it, so it is preferable to have a separate slide of magenta colour.

Scarlet—if a magenta slide is available, the combined use of magenta and red produce the scarlet. If a magenta slide is not available, a slightly different shade of scarlet results from combined use of blue and red.

Purple—if a magenta slide is available, the combined use of magenta and violet produces the purple. If a magenta slide is not available, the combined use of yellow and violet produces a slightly different shade of purple.

Chapter 20   CONDENSED SUMMARY OF THE ATTRIBUTES OF THE TWELVE COLOURS
(After Dinshah Ghadiali)

RED
  Sensory stimulant
  Liver energizer
  Irritant, vesicant, pustulant
  Rubefacient, caustic
  Haemoglobin builder

ORANGE
  Respiratory stimulant
  Para-thyroid depressant
  Thyroid energizer
  Anti-spasmodic
  Galactagogue
  Anti-rachitic
  Emetic (under conditions mentioned
    in text)
  Carminative, stomachic, aromatic
  Lung builder

YELLOW
  Motor stimulant
  Alimentary tract energizer
  Lymphatic activator
  Splenic depressant
  Digestant
  Cathartic
  Cholegogue
  Anthelmintic
  Nerve builder

LEMON
  Cerebral stimulant
  Thymus activator
  Antacid
  Chronic alterative
  Anti-scorbutic
  Laxative
  Expectorant
  Bone builder

GREEN
  Emotional stabilizer
  Pituitary stimulant
  Disinfectant, antiseptic
  Germicide, bactericide
  Muscle and tissue builder

TURQUOISE
  Cerebral depressant
  Acute alterative
  Acid, tonic
  Skin builder

BLUE
  Anti-pruritic
  Diaphoretic
  Febrifuge
  Anodyne
  Demulcent
  Vitality builder

INDIGO
  Para-thyroid stimulant
  Thyroid depressant
  Respiratory depressant
  Astringent
  Sedative
  Pain reliever
  Haemostatic
  Inspissator
  Phagocyte builder

VIOLET
  Splenic stimulant
  Cardiac depressant
  Lymphatic depressant
  Motor depressant
  Leucocyte builder

SCARLET
  Arterial stimulant
  Renal energizer
  Genital excitant
  Emmenagogue
  Vasoconstrictor
  Ecbolic
  Sex builder in those with sub-
    normal sex potency

MAGENTA
  Cardiac energizer
  Supra-renal stimulant

  Diuretic
  Emotional Equilibrator
  Auric builder

PURPLE
  Venous stimulant
  Renal depressant
  Anti-malarial
  Vasodilator
  Narcotic, hypnotic
  Analgesic, anti-pyretic
  Sex sedative

Warning Regarding Self-diagnosis and Treatment

As chromo-therapy is a method within reach of the layman, we feel it is necessary to remind our readers that for any major ailment it is always dangerous for laymen to try to do their own diagnosing and treating.

The material on chromo-therapy is included in this volume partly for the light it throws on what appears to be an important set of factors influencing the state of health or illness in the body, and partly for the information of doctors who did not happen to come in contact with the courses given on chromo-therapy in the past, but who are interested in natural methods of healing and may want to add this therapy to their other methods of treatment.

In practice, laymen naturally do not run to a doctor every time they have a minor ache or cold—so a certain amount of self-diagnosing and treatment does occur. The vast sales of drugs, vitamins, etc. to laymen without prescriptions is ample evidence of this. To the extent that laymen inevitably will resort to self-diagnosis and treatment for minor ailments, the section of this book on chromo-therapy may be of interest to them.

Chapter 21   FACTORS INVOLVED IN PRODUCTION OF THE SLIDES

The shade of colour incorporated in the slide is important, as best results in chromo-therapy are produced when the frequency of light transmitted by each slide is in the middle of the frequency range of the colour used. Also, since some of the colours are produced by super-imposition of various combinations of two slides, a deviation in shade (and therefore of vibratory frequency transmission) of any one of the slides is going to affect not only its own colour but also the frequency transmitted by any combination in which the slide in question is one of the members. Data is available on the exact peak frequency of light which each of the slides should transmit. However, it is most difficult in these times to find a supplier who can furnish coloured glass produced in accordance with frequency specifications. At one time in the past, very excellent glass of this nature was obtainable in Czechoslovakia, but that was before that country fell under foreign domination.

Not only is there a difference in peak frequency transmitted by a coloured-glass slide, but there is also a difference in width of the frequency band that is transmitted. Some who have done research in colour-therapy, advocate a relatively narrow band of transmitted frequencies.

There is a controversy in the field as to whether the frequencies transmitted by a slide constitute the only factor in the effect on the living organism, or whether the chemicals which go into the production of the glass have an effect apart from the frequencies transmitted, i.e. it could be possible to have the same shade of colour through use of different chemical mixtures, and one school of thought claims that each mixture will have a slightly different effect. Babbitt, in his *Principles of Light and Color* gives some information on what he believed to be the varying effects of colours produced from different chemicals incorporated in the production of glass.

In the absence of glass made specially in order in accordance with frequency transmission characteristics, about the best that can be done is to match, from the shades available from importers, as closely as possible the shades of colours used in the therapeutic colour lamps manufactured in the nineteen-thirties in the U.S.A.

Chapter 22   OTHER METHODS OF CHROMO-THERAPY

Besides the systemic and local irradiation of the body with the rays of coloured light, there are several other methods of administering the effects of the colour vibration.

1. *The use of colour rays to "charge" water*, which is then ingested. In this application, water is placed in a glass container of the desired colour, and exposed to sunlight for a period of time. The water then takes on the attributes of the colour of the glass container, i.e. water charged in yellow glass bottles or jars has a laxative effect; water charged in green glass has a cleansing, bactericidal effect, etc.

2. *The use of colour rays to charge sugar of milk*. By irradiating milk sugar with the desired colour ray, the powder can be taken into the system repeatedly, in small quantities, to convey the resulting colour vibrations.

3. *The use of coloured lenses*, through which to look at a bright light. In this application, termed by some as "Thalamotherapy", lenses are placed over the eyes, of the colour desired, while a bright source of light is placed a few feet away. The effect is said to be engendered through the medium of the optic thalamus.

Each of the above methods can be applicable to some extent when a systemic effect is desired, but they lack the specific local effect obtainable from Ghadiali's method of applying certain colour rays to the area of the organs or the pathological locations which it is desired to treat. There are cases where internal consumption of colour-charged water or sugar is helpful in supplementing the rays beamed to the body and its aura.

So far as we know, there has been only one improvement made by others over the work of Dinshah Ghadiali. A firm formerly operating in Chicago under the name of Actino Laboratories, inaugurated the use of:

4. *Separate shades of blue and yellow* for the two opposing physical types. In this application, the stout, sluggish individuals are termed the anabolic type, and the thin, nervous individuals are termed the catabolic type. For the catabolic individuals, a shade of blue somewhat nearer the indigo is used, in order to exert a more calming effect. For the anabolic individuals, a shade of yellow somewhat nearer the orange is used, in order to exert a more stimulating effect. Thus in this method there are two different blues from which to choose, and two different shades of yellow.

APPENDIX I. CHROMO-THERAPY

Correlation between Elements and Colours

Following is a table showing the colour that was found to be *predominant* in spectroscopic analysis of each of the elements listed, as determined by extensive research performed by Dinshah Ghadiali with elaborate physics and optical equipment. The task was complicated by the fact that each element has a number of fundamental lines.

Methods of spectroscopic analysis used in this project included disintegration in alcohol or bunsen burner flames; in high-tension spark; in vacuum tubes; and in burning in electric carbon arc with carbon lines removed.

RED: Cadmium, *hydrogen*, krypton, neon.

ORANGE: Aluminium, antimony, arsenic, boron, *calcium, copper*, helium, selenium, silicon, xenon.

YELLOW: *Carbon*, glucinum, iridium, magnesium, molybdenum, osmium, palladium, platinum, rhenium, *sodium*, tin, tungsten.

LEMON: Cerium, germanium, *gold*, hafnium, *iodine, iron*, lanthanum, neodymium, phosphorus, praseodymium, protoactinium, ruthenium, samarium, scandium, *silver, sulphur*, thorium, titanium, uranium, vanadium, yttrium, zirconium.

GREEN: Barium, *chlorine*, kasmirium, *nitrogen*, radium, tellurium, thallium.

TURQUOISE: chromium, columbium, *fluorine, mercury, nickel*, tantalum, *zinc*.

BLUE: Caesium, indium, *oxygen*.

INDIGO: *Bismuth*, ionium, lead, polonium.

Violet: Actinium, *cobalt*, gallium, niton.

Scarlet: Argon, dysprosium, erbium, holmium, lutecium, *manganese*, thulium, ytterbium.

Magenta: Irenium, lithium, *potassium*, rubidium, strontium.

Purple: *Bromine*, europium, gadolinium, terbium.

Appendix II. Chromo-Therapy

Red, green, and Violet as the Primary Colours of Light

1. The visible spectrum at the high frequency end, starts not at Blue but with Violet. Green is the middle colour in the visible sequence of Violet, Indigo, Blue, Green, Yellow, Orange, Red.

2. If green light was the product of yellow and blue and therefore a secondary colour, then green could not continue to exist when yellow and blue have disappeared. Turn a prism so that all seven colours are seen, then gradually revolve the prism—yellow and blue disappear and only red, green and violet are left. Any further rotation eliminates all colour until the full spectrum again re-appears.

3. If yellow light was a primary colour, it could not be a product of other colours. But red and green lights projected from two lanterns on a screen, merge into yellow.

4. Similarly, violet and green slides produce blue when the beams merge on a screen, showing that blue cannot be a primary colour.

5. In a darkened room, look through a prism at an electric bulb with a rheostat in the circuit. Start with no current—as rheostat is turned, red appears first; with more light, green becomes visible; with still more light, violet makes its appearance, and with additional light, all seven colours show.

ULTRA-RED RAY THERAPY

Summarized from Dr. George Starr White, M.D.

Dr. White became interested in the real science behind ultra-red therapy, when he observed that the heat radiated by certain of the heavy-metal cook-stoves (the old-fashioned kind) was beneficial to cases of arthritis, neuritis, etc., while heat radiated by cook-stoves made from other metals was harmful and irritating. This led him to make an intensive investigation into the fundamental factors involved in what he terms "radiant light therapy". He also coined the term "Ultra-red ray therapy", to differentiate it from conventional infra-red therapy which makes no distinction between the irritating and non-irritating rays which lie in the frequency band just below the visible light band.

In his investigations, Dr. White developed special laboratory instruments which he states detect energy after it has been altered by being passed through the human body. With these instruments he found:

1. That a change takes place in the character of the energy of light through being passed through the organism. (And we here use "light" in its broader sense of radiant energy whether in the visible band or not.)

2. That the energy passed into the body is not the same as the energy that contacts the skin—skin is a natural filter and produces changes.

3. That the energy resultant passing into the body after being filtered through the skin, is to some extent absorbed by the body fluids; the extent of absorption depending on the proportion of irritating and non-irritating components—it is the latter that are absorbed.

4. That certain widely-held conclusions as to the therapeutic properties of radiant energy of various frequencies are found to be incorrect, as they do not take into consideration the characteristics of the absorbed "energy resultant".

For example, Dr. White states that quartz and mercury vapour lamps produce an unbalanced ray spectrum, which does not produce nearly as good results as his methods; in fact they can produce congestion in deeper structures. One of the main principles discovered by Dr. White in this work, is that

> THE RADIANT ENERGY RESULTANT that is absorbed internally, DIFFERS WITH THE SOURCE OF THE ENERGY, AND ALSO DIFFERS WITH THE TYPE OF FILTER USED.

Says Dr. White: "Light is a natural food, and radiant energy is a catalytic agent, causing the body to take from other food that which it would not obtain were the radiant energy not administered."

He found that the skin pores close up when subjected to a steady flow of strong light or radiation of heat, and that this shutting off, greatly reduces the ability of the body to absorb radiant energy. This shutting off is avoided and far better results obtained, if, through a motor-driven rheostat mechanism, the light source is constantly faded in and out. Both the ratio of the "fading in" time to the "fading out" time, and the duration of the total cycle, is important from the standpoint of results obtained.

Added benefits were obtained by correlating the fading in and fading out, with the patient's breathing—by having the lamps fade in during inhalation, and fade out during exhalation. . . . The best ratio was found to be 4:5—4 for inhalation with the light fading in, 5 for exhalation with the light fading out. The patient is instructed to match his breathing with the variation in light. The greatest benefit was found to be obtained during periods when the lights fade out. During those periods, a greater proportion of the radiant energy is absorbed by the body, as proven by measurements taken with the devices for detecting energy after it has passed through the body. During those periods, deeply penetrating ultra-red rays produce *conversive heat* internally, and relieve congestion and inflammation without injuring the patient. Circulation is enhanced through relieving congestion. The ultra-red rays are the best known pain-relieving agents and also normalize blood pressure, and stabilize the vaso-motor mechanism, according to Dr. White's findings in his very extensive clinical practice which he maintained for many decades.

The unit controlling the duration of the fading cycle should be adjustable; at first it should be set to coincide with the respiratory rate of the patient—then day by day the duration of the cycle should be lengthened, up to a maximum of thirty seconds for one complete cycle of inhalation and exhalation.

SOURCE OF RADIANT ENERGY

Dr. White spent many years of research with the ultra-red rays. He found that their generation was easy to induce, since any metal object when heated would emit these rays, but far more important, he discovered there was a vast difference in the therapeutic properties in the rays emitted by different substances and at different temperatures! It was this discovery which enabled him to achieve results far superior to others using rays from heated objects. Dr. White found that the rays from conventional heat lamps (infra-red ray lamps) are far too irritating for best results.

Dr. White's work in this field showed the need for attention to three factors in the production of the most beneficial ultra-red rays:

1. The type and size of heating element.

2. Using a reduced amperage at steady flow for the light source used over the head, and the fading cycle already described, for the lights used over the body.

3. The type of screen or filter interposed between the lights and the body.

Considering each of these in turn:

1. Passing over Dr. White's list of the many substances he investigated, which he found when heated, to emit rays of an irritating character, we come to his determination of the apparatus which produces the most beneficial type of ultra-red rays (and he used the term "ultra-red" rather than "infra-red" advisedly)—

Namely—a 1,500-watt tungsten electric light in a clear globe should be used, in well-ventilated reflectors that spread the light evenly, giving practically no foci.

2. The lamp to irradiate the head should definitely be operated *below* its rated voltage and amperage, if a high quality of therapeutic ray is to be at-

tained. This is accomplished by having an adjustable, heavy-duty resistance in series with the lamp. The resistance to be adjusted in accordance with the complexion and skin colour of the patient. Dr. White found that blondes received the maximum "energy resultant" when the current to the lamp was reduced from the rated 13 amps to the range of 9–10 amps, while negroes received the maximum energy resultant at 4 amps for darkest skin—5 amps for skin less dark. Between these extremes, there is an exact amperage that gives the greatest penetration, or energy resultant, for each type of person. Brunettes should receive 7 to 9 amperes. It is important to note that a smaller lamp operated at rated amperage and voltage does not give the same result as a larger lamp operating at a lower than rated amperage as specified in this paragraph!

When the rays for the head are properly reduced, they can fall either on the front or back of the head.

3. Dr. White found that a screen below the bulb in each reflector served not only as protection against touching the hot bulb and against bulb breakage, but also increases the amount of ultra-red rays and renders the radiant energy more easily absorbed through to the skin. These benefits varied according to the type of screen used.

(*a*) Best material was found to be phosphor-bronze; copper is next.

(*b*) Therapeutic benefit is considerably greater from a double screen rather than a single one.

(*c*) The best results were obtained when the screens were positioned so that the meshes cross each other, and are $\frac{3}{8}$ in. apart at the area of maximum light passage.

A FEW NOTES ON THE USE OF ULTRA-RED RAY THERAPY

By using several large reflectors in a row, the patient, lying on the treatment table, gets better treatment and faster results. At least three reflectors are advocated—one for the head and two for the body. The lamp globes, states Dr. White, should be 12 in. to 15 in. from the areas to be treated, and should radiate squarely at the skin surface. He states that first the back should be radiated for ten minutes, to relax the muscles around the spinal column, then radiate the front of the body for thirty minutes.

In the treatment of tuberculosis of the lungs, Dr. White recommends that the lamp over the head and lungs be set for steady raying for twenty minutes, then let it fade in and out with the other two lamps.

For bone TB, he used the reduced amperage of light steadily over the affected area of bone for the first part of the treatment—for the second part, it faded in and out with the other two lamps.

INVISIBLE ULTRA-VIOLET LIFE FREQUENCIES
MADE VISIBLE

Microscopes constructed on entirely new principles have been invented and developed by Royal Raymond Rife, a scientist and biologist. These new microscopes have led to many discoveries regarding:

(*a*) The characteristics of bacteria and virus micro-organisms.

(*b*) Factors leading to the production or transformation of various micro-organisms.

(*c*) The role of bio-chemical changes, in encouraging or retarding the growth of harmful micro-organisms.

In this section we shall first consider briefly the new features of the Rife Microscopes, and then summarize the discoveries that have occurred through the use of these instruments.

1. THE RIFE MICROSCOPES

With these outstanding optical devices, resolution up to 31,000 diameters and magnification up to 60,000 diameters is obtained, with a number of advantages over the electron microscopes—the only other devices known at the present time which reach such high magnifications.

The results obtained from the Rife microscopes are due principally to the use of three principles of physics in a manner completely new to the field of optics:

1. A method of selecting a portion of the frequency spectrum of light for use in viewing specimens.

2. A method of heterodyning light to bring micro-organisms of various invisible ultra-violet colours into the visible light frequency range.

3. The attainment of very high magnification and resolution through an ingenious method for keeping the optical rays parallel in the instrument.

Considering each of the foregoing principles in turn:

1. *Selecting a Portion of the Frequency Spectrum*

It is well known that a beam of light passed through a prism is broken up into the colour spectrum, and since different colour-components of the beam are displaced by differing degrees, the beam emerging from the prism is spread out over a relatively wide area. The visible colours can be seen un-aided, but beyond the red component of the beam there is an invisible beam of infra-red and beyond the violet component there is an invisible area of ultra-violet frequencies transmitted by the prism if it is made of material such as quartz which permits the transmission of ultra-violet.

In the Rife microscopes, circular, wedge-shaped, block-crystal quartz prisms are used to polarize the light to be sent through the scope. By means of a revolving adjustment or control, the portion of the spectrum sent through the prisms is selectable, so that a narrow band corresponding to any colour from infra-red up through the visible colours and then through the entire ultra-violet range in narrow steps, can be selected for use in illuminating the specimens. The importance of this unique feature will be evident later.

2. *Heterodyning Light*

We will first explain the term "heterodyne" and then show its application to light as developed by Rife. It is an observed fact in physics, and a principle constantly used in radio and in work with sound, that when two different frequencies of vibration are produced, they inter-act upon each other to produce two new frequencies—one of which is the *sum* of the two original or fundamental frequencies; the other is the *difference* between the two originating or fundamental frequencies. Suppose, for example, that in the range of sound, a tone of 400 cycles per second and another tone of 600 cycles per second is produced. The resulting new frequencies will then be 200 cycles (the difference between 400 and 600 cycles—and 1,000 cycles for the other new tone, the sum of 400 and 600 cycles.

So far as is known, Rife was the first individual to apply this principle to the field of light. The visible frequencies range from about 436 trillion oscillations per second at the red end of the visible spectrum, to about 732 trillion oscillations per second at the violet end of the visible spectrum. An oscillatory rate faster than 732 trillion times per second results in a beam which is in the invisible, ultra-violet range. The ultra-violet band occupies several octaves of vibration, as compared to the visible spectrum which occupies less than one octave of vibration. . . . (The upper limit of an octave has twice the vibratory rate of the lower limit of the same octave.) So the range of the vibratory light spectrum invisible to the human eye is larger than the frequency range of the light spectrum which the eye can perceive.

The process of heterodyning light is accomplished by bringing an invisible, ultra-violet beam of, for example, 1,200 trillion oscillations per second into contact with another equally-invisible beam of say, 1,700 trillion oscillations per second; the difference between the oscillatory rates of the two originating beams results in the production of a light beam having an oscillatory rate of 500 trillions per second, which is within the range visible to the human eye.

In the past, many micro-organisms could only be observed if stained with a chemical. Some micro-organisms never became visible with other microscopes, because no suitable stain could be found for them. One of the prime advantages of the Rife microscopes is that Rife found many of the micro-organisms having no colour in the visible light range—their frequency characteristic is such that they have a "colour" in the invisible, ultra-violet range. By the use of the heterodyning principle in his microscopes as mentioned, the micro-organisms of ultra-violet colours are brought into the visible light range in their natural state, without the use of any stain. This method also brings into visibility the micro-organisms which had not responded to any known stain, and all micro-organisms can be viewed in their natural live state—a very considerable advantage, since the use of a stain kills the micro-organism. In fact this is the only microscope yet known by which ultra-high magnification can be used to view organisms in their living state, for the beams from electron microscopes instantly kill any living organisms.

3. *Achievement of Very High Magnification through Optical Means*

In the ordinary microscope, the rays of light refracted by the specimen enter the objective and are then carried up the tube in supposedly parallel rays, but in practice these rays converge after a certain distance, cross each other, and then diverge, resulting in distortion and a limit on the amount of magnification obtainable, since the rays by ordinary means cannot be kept parallel for a sufficient distance to pass them through several series of lenses. In the Rife microscopes, specially-designed quartz prisms are inserted into the tube at frequent intervals to counteract the tendency of the rays to diverge from parallel. This enables three matched pairs of oculars to be used in the universal microscope, the largest which Rife has constructed, permitting the attainment of the extraordinarily high powers of magnification and resolution that we have already mentioned. The supposed limit on magnification arising from the dimension of a wavelength of the light used for viewing the specimen, has been transcended by Rife, partly through the utilization of ultra-violet light which is composed of wavelengths of shorter dimensions than those of visible light, and partly by other means. Many technical details of the instrument are contained in the article *The New Microscopes* by R. E. Seidel, M.D. and M. Elizabeth Winter, published in the February, 1944, Journal of the Franklin Institute. That article has recently been reprinted by the Lee Foundation for Nutritional Research, Milwaukee 3, Wisconsin, and published as their Reprint #47.

2. DISCOVERIES RESULTING FROM THE USE OF THE RIFE MICROSCOPES

1. *Frequency Characteristics of Micro-organisms*

The adjustment or control mechanism in the Rife microscopes, for selecting the frequency band of light sent up through the lenses, has already been mentioned. In the use of these instruments, it is found that the control setting differs for every different type of bacteria and virus, and that for any particular type of bacteria or virus the setting is always the same. This means that each different type of bacteria and virus has its own characteristic life frequency which it emits, and by "tuning" the microscope to that frequency of light, the micro-organism becomes brilliantly visible without the use of any chemical stain.

In the use of the Rife microscopes it has been found, for example, that Bacillus Typhosus is always a turquoise-blue; Bacillus Coli is mahogany-coloured; Mycobacterium Liprae is always a ruby shade; the filter-passing form

T.E.—E

of virus of tuberculosis is an emerald green; the virus of cancer (one of the discoveries made possible by the Rife microscopes) is purplish-red, etc. Different colours are of course representative of different frequencies of light.

2. Observations of Micro-organisms not shown by Other Microscopes

Because of the unique characteristics of the Rife instruments as already described, they permit observation of micro-organisms which other microscopes are unable to show. Among the discoveries thus made, have been virus organisms present in poliomyelitis and cancer.

3. New information regarding the relationship between Micro-organisms and Their Chemical Environment

The cancer virus which was isolated by Rife, and which he terms BX virus, induced cancer growths in 104 successive generations of albino rats. During the course of the extensive experiments performed with this virus, it was found that with a slight change in the chemical media for the culture, a larger virus resulted, termed BY. Another slight change in the chemical media, and the virus is transformed into a monocyte. With still another change in the chemical environment, the monocyte becomes a fungi, and with a still further slight change, the fungi turns into Bacillus Coli! Then if the Bacillus Coli is kept in a certain media for a year (the time required for metasteses), the BX virus again appears! The changes in the chemical environment required to effect these transformations are very slight—in fact it is stated that an alteration of four parts per million in the media will transform the harmless B. Coli into the deadly B. Typhosus. These changes can be made to occur in as short a period as forty-eight hours.

It is Rife's belief that all pathogenic (disease-producing) micro-organisms are divided into ten groups, and that any micro-organism can be converted into that of any other within its group, by changing the chemical environment, sometimes by as little as two parts per million. From the above it can be seen how slight metabolic changes in body tissues can induce a micro-organism of one group to change into another micro-organism within the same group. The Rife work provides interesting support for, and visual confirmation of, the Naturopathic theory. In contrast to the Allopathic view, Naturopaths hold that the important factor in fighting disease is the vitality of the patient and the strength of the general constitution, and that if these can be supported and the body chemistry kept balanced, germs need not be a concern.

4. Use of Selected Frequencies of Radiation to Destroy Specific Micro-organisms

To quote from the article in the Journal of the Franklin Institute:

> "Under the Universal (Rife) Microscope, disease organisms such as those of tuberculosis, cancer, sarcoma, streptococcus, typhoid, staphylococcus, leprosy, hoof and mouth disease, and others may be observed to succumb when exposed to certain lethal frequencies peculiar to each individual organism, and directed upon them by rays."

The frequencies referred to in the above paragraph are in the radiowave band, and the most effective method of administration has been found to be the use of these differing frequencies of radio waves to pulse the current of a vacuum tube similar to an X-ray tube but partially filled with helium, so that none of the destructive X-ray radiations are emitted. The beam or rays from this new type of tube is directed at the micro-organisms under consideration. This work is in the laboratory stage, and is of interest mainly because of the principles involved.

Once it was proven, by the use of the Rife microscopes, that each type of micro-organism has its own particular life frequency or rate of vibration in the light band, it became a logical corollary that for each type of micro-organism there is also some frequency radiation or rate of oscillation that will be destructive to the organism.

In the field of radionics, for example, the theory has long been maintained that each virus, bacteria or type of toxin has its own frequency of radiation or tuning, and that these frequencies provided a key to tunings which could be used to destroy the virus or bacteria—however with Rife's work it is now possible to prove the correctness of the theory, by observation with his special microscopes which show the destruction of any micro-organism when the appropriate frequency of radiation is applied.

THE ORGONE

THE ORGONE

INTRODUCTION

SOME of the therapeutic methods in this book are presented by their proponents, solely on the basis of empirical results. With others, varying amounts of scientific research have been undertaken to discover the underlying hypotheses and to establish the relationship between the methods, and the basic principles of science. Of all the methods discussed in this volume, the most extensive research job has been accomplished with the orgone energy discovered by Dr. Wilhelm Reich, M.D. He has been working with this energy for sixteen years, much of the time assisted by a sizeable staff. The resulting discoveries have the most profound implications in many different fields of human knowledge. In fact, they affect some of the most fundamental principles of the following subjects:

Physics
Biology
Bio-Genesis (production of living organisms from non-living matter)
Bacteriology
Pathology
Haematology (study of the blood)
Psychiatry

The literature already published by the Orgone Institute Press includes several large volumes and many detailed bulletins issued over a ten-year period in the United States of America. It is impossible to do full justice to the subject of orgone energy in any one book, to say nothing of part of a book—which is all that can be devoted to the subject in the present instance. In this volume, we try to give the reader a condensed summary of some of the principles of the orgone energy, but we cannot undertake adequate proof of the revolutionary theories propounded by Dr. Reich. The reader is referred, therefore, to the literature published by the Orgone Institute Press, which gives detailed accounts of numerous supporting experiments.

1. BIO-GENESIS

In *The Discovery of the Orgone*, Volume II, 1947, Chapter Two, Dr. Wilhelm Reich discloses in detail the method he discovered for transforming particles of inanimate matter into living particles that pulsate, have internal movements of their contents, and themselves move as a unit from one place to another, and gain the ability to react to a biological stain. He terms these particles "bions". The bion is regarded as the transitional form, from non-living to living matter.

In brief, the process of transforming particles of non-living, inert matter into bions is accomplished by autoclaving a solution of 50% bouillon and 50% Kc (potassium chloride) solution, and adding to that solution, dust from coal or other substances after that dust has first been heated in a gas flame to white incandescence. The process must be done under certain conditions which Dr. Reich and his staff have developed in detail, for the formation of a colloidal solution.

The result, viewed microscopically, as described by Dr. Reich follows: "At 2–3000× ... we notice that small vesicles of about one micron diameter detach themselves from the margin of the larger particles and move about freely in the field. In successful preparations, one notices motility at the margins of the particles in the form of expansion, contraction and vibration. If we observe long enough, we see how the small particles undergo a change before our very eyes. At first they appear "rigid", with a thick and black membrane. Gradually, the membrane becomes thinner. On the inside of the particle, we notice increasingly a blue or blue-green shimmer. The vesicles become tauter and show more motility inside. Many of them show an undulating vibration of their contents. The thinner the membrane becomes, the more intense becomes the blue glimmer and the more vivid the motions. Soon—the same day, but more distinctly the following day—we see movements of expansion and contraction. No one who has closely studied these preparations has had the slightest doubt as to the living character of the movements. One can distinguish movements from place to place and internal movements of the contents, displacement of the blue colour and of the intensity of the light phenomena, protrusions and retractions of the content; the vesicle pulsates in an irregular rhythm."

The bion vesicles that are mobile are found to be positively charged, since they move towards the cathode of a galvanic current. Later, when the vesicles have lost their motility and are no longer in colloidal suspension, they are found to have lost their positive charge.

In the use of biological stains, control experiments using unprepared coal dust show that the carbon particles remain black and do not react to a Gram or carbol fuchsin stain. On the other hand, coal *bions*, produced according to the process described, immediately react positively by turning blue with the gram stain. Quoting Dr. Reich: "It can be observed, furthermore, that only those particles react which have already undergone a certain bionous development (thin membrane, increased fluid and blue colour inside) while the undeveloped particles react neutrally like those in the control preparation."

The early experiments with bions were made with carbon, but it was found that carbon was by no means the only substance that would produce bions. In fact, Dr. Reich found that the more substances he examined, the more inevitable became the conclusion that all matter, if heated to incandescence and made to swell, consists of or is transformed into the blue vesicles which he terms bions. Various examples are given in detail in the book previously cited.

Some solutions, as for example one described that is composed of certain proportions of gelatine, egg white and lecithin in water, filtrated bouillon and O.Ln Kcl, produce not only individual bions but also collections of blue vesicles which merge into a "heap" that expands or contracts of itself—termed by Dr. Reich an ameboid.

Movement of the bions from one place to another (external movement) ceases when the bions no longer have internal movement, i.e. pulsation, expansion, contraction, vibration and glimmering. Without going into the reasons and proof, we will mention here Dr. Reich's statement that the internal motility is the result of an internal charge of energy, and that this biological energy which produces the internal impulses derives from the very same matter which composes the bion.

2.　DERIVATION OF THE TERM "ORGONE"

Dr. Reich was a psychiatrist. It was through his investigation of certain emotional functions, particularly referring to the nature of the sex mechanism and the characteristics of the orgasm, that he was led into the laboratory experimentation which produced the vast new structure of knowledge which is now included under the heading of orgonomy. As our present book deals with certain aspects of energy manifestations, it is not our purpose to become involved in theories that pertain to other fields. We shall mention briefly Dr. Reich's theory of the relation between orgone energy and the orgasm, at the same time differentiating the theory from factual results of specific experiments.

Quoting from Dr. Reich, "... The orgasm is a fundamental biological phenomenon; fundamental because the orgastic energy discharge takes place in the form of an involuntary contraction and expansion of the total plasma system. Like the respiratory function, it is a basic function of any animal system. Biophysically speaking, it is impossible to distinguish the total contraction of an ameba from the orgastic contraction of a multi-cellular organism. The outstanding phenomena are intensive biological excitation, repeated expansion and contraction, ejaculation of body fluids in the contraction, and rapid reduction of the biological excitation ... In its quickly alternating expansions and contractions, the orgasm shows a function which is composed of tension and relaxation, charge and discharge: *biological pulsation.*

"Closer investigation reveals the fact that these four functions appear in a definite four-beat: the mechanical tension, which shows itself as sexual excitation, is followed by a bio-energetic charge of the periphery of the organism. This fact was unequivocally demonstrated by measuring the bio-electric potentials occurring with pleasurable excitation of the erogenous zones. When tension and charge have reached a certain degree, there occur contractions of the total biological system. The high peripheral charge of the organism is discharged. This is seen objectively in the rapid decrease of the bio-electric skin potential; it is felt subjectively as a rapid decrease in excitation.

"To us, the function of the orgasm is expressed in the four-beat; mechanical tension, bio-energetic charge, bio-energetic discharge, and mechanical relaxation. We shall call it the function of tension and charge. We know from earlier investigations that the function of tension and charge is characteristic not only of the orgasm. It applies to all functions of the autonomic life system. The heart, the intestines, the urinary bladder, the lungs, all function according to this rhythm. Cell division also follows this four-beat. So does the movement of protozoa as well as metazoa.

"Thus it is obvious that there is one basic law which governs the organism as a whole as well as its autonomic organs. The total organism contracts in the orgasm just as does the heart with every pulse beat; the medusa as a whole contracts just as does the vacuole in a protozoon. This biological basic formula comprehends the essence of living functioning. The orgasm formula shows itself to be the life formula as such."

From the above quotation, showing Dr. Reich's theories on the relation between the orgasm and the life processes, it will be evident why he chose the term "orgone", which, as he states, is derived from the words *organism* and *orgastic*. To quote again, "From now on, the term 'orgonotic' comprises all energetic phenomena and processes which specifically pertain to the energy which governs living matter. Every living organism is a membranous structure which contains in its body fluids an amount of orgone; it is an 'orgonotic' system."

3. RADIATION FROM BIONS

Dr. Reich and his staff produced bions from many different materials, as reported in his book *The Discovery of the Orgone*, Part II. One of the most surprising developments of this research was the discovery that they could be produced from ordinary beach *sand*, which was heated, sterilized, and autoclaved (kept in a steam chamber under pressure) in the regular culture medium of bouillon and potassium chloride. The resulting growth, when inoculated on egg medium and agar, resulted in large, slightly mobile, intensely blue packets of energy vesicles. This culture consisted of just one kind of form. Examination under a microscope with magnification of 2–4,000 × showed forms which refracted light strongly, consisting of packets of six to ten vesicles and measured about ten to fifteen microns. These bions were termed SAPA (from the words *sand* and *packet*). They showed characteristics of unusual interest.

When brought near cancer cells, the SAPA bions killed or paralyzed those cells even at a distance of ten microns. When cancer cells came as close as that to these bions, they remained in one spot, as if paralyzed; they would turn around and around in the same spot and finally become immobile. Dr. Reich recorded this phenomenon on microfilm.

It was found that daily observation of the SAPA bions through the microscope resulted in eye irritations and conjunctivitis of the eyelids, to a far greater extent than occurred from the same duration of microscope use with other objects. This was the first effect noted, of what was later termed "orgone radiation". Attempts to identify the nature of this radiation and discover its characteristics, at first met with many failures.—"The customary methods of radiation research gave no results. The orgone radiation required the elaboration of special hitherto unknown methods and apparatus which could be achieved only step by step, by long-continued observation."

It was then found that SAPA bion cultures, placed on a glass slide held in the palm of the hand, produced discoloration of the skin, and after repeated application, an inflamed and painful condition of the palm.

The air in the room where the cultures were kept became extremely "heavy", and people who stayed in the room developed headaches if the windows were closed for as short a time as one hour.

Radiation from the cultures were found to register on photographic plates kept in the room in the dark, but control plates kept in the same room were also fogged, apparently leading to the conclusion that the "energy" involved was everywhere in the room.

Peculiar optical phenomena developed when dozens of the SAPA bion cultures were prepared and kept in the room simultaneously.

"After the eyes became adapted to the darkness, the room did not appear black, but grey-blue. There were fog-like formations and bluish dots and lines of light. Violet light phenomena seemed to emanate from the walls as well as from various objects in the room. A magnifying glass held before the eyes made these impressions more intense, and the individual lines and dots became larger, showing that they had an objective reality (since a magnifying glass will not magnify subjective phenomena occurring internally in the observer)." The energy involved had an irritating effect on the optic nerve, which produced after-images that persisted after the eyes were closed.

With the intensified amount of this odd energy in the dark room, the eyes of an observer would become red and painful even when no microscope was used, but simply from the effect of being in the room for an hour or two. More prolonged exposure resulted in a curiously luminous phenomenon in which one could see a radiation from one's palm, shirt sleeves and, with a mirror, from one's hair. The blue glimmer would then be visible as a slowly-moving, grey-blue vapour around one's body and around objects in the room. People who were told nothing about these effects, observed exactly the same phenomena when placed in the room in the dark. The whole situation is reminiscent of a fantastic science-fiction story, yet was proven time and again by scientifically-controlled experiments.

One of the devices which reacts to radiation of radium, static electricity and similar influences is the electroscope. In this device, pieces of gold leaf or aluminum leaf are suspended in a glass case, and are repelled or attracted by positively and negatively charged radiation. . . . When tested with the radiation from the SAPA bion cultures, there was no reaction on the gold leaf in the electroscope. Later it was found that rubber gloves, which had been left near the SAPA cultures for a period of time had absorbed radiational energy from those cultures and would produce a very strong reaction in the electroscope when brought near that device. If the gloves, after exposure to the culture, were aired outside the building for fifteen minutes or more, they became neutral and did not influence the electroscope. It was then found that paper, cotton, cellulose and other *organic* insulating substances took up an energy from the SAPA bion cultures which gave a reaction in the electroscope. High humidity, ventilation in the shade, or touching the substances with the hands for several minutes, all eliminated the effect.

The fact that the radiation from the cultures would affect an electroscope only through the indirect route of being used first to "charge" an insulating material, was evidence that the radiation was of a hitherto unknown type, with different characteristics from any then known to science.

Further experiments showed that new rubber gloves would also affect the electroscope, although not so strongly as those which had been "charged" by being placed near the cultures. Placing new rubber gloves in the sun for from five to fifteen minutes, resulted in a very strong reaction when these gloves were then brought near the electroscope. The import of these experiments appeared to be that a certain amount of this type of energy was everywhere, and that its origin was from the sun. Was it possible that the ocean sand had absorbed sun energy, and that the process of transforming the sand into SAPA bions releases this particular type of sun energy? Dr. Reich thinks that is exactly what takes place. As for us, we shall be content to report the results of his experiments, which surely are epoch-making in themselves, and leave the theorizing to others.

4.　OBSERVATION OF ORGONE ENERGY IN THE ATMOSPHERE

As Dr. Reich states, "In order to study the radiation from the SAPA bions, a closed space had to be constructed which would close in the radiation and prevent it from rapid diffusion into the surroundings. No organic material could be used for this purpose, since, as we have seen, organic material absorbs this energy. According to my observations, metal, on the other hand, would reflect the energy and confine it within the enclosed space. However, the metal would reflect the energy also to the outside. In order to avoid this, the apparatus had to have metal walls on the inside, and walls of organic material on the outside. With this construction, it was to be expected that the radiations from the cultures would be reflected by the inner metal walls, while the outer layer of organic material (cotton or wood) would prevent or at least reduce the reflections to the outside. The front wall of the apparatus was to have an opening with a lens through which the energy could be observed from the outside."

In order to assure the visibility of the orgone rays, it was decided to use a disc of organic material in conjunction with the lens system. A cellulose disc in a film-type viewer was used. What we have now is a box with metal as an inside wall and wood on the outside, and an aperture in which the lens was inserted in a tube. Using this observation box, it was found that the radiations from SAPA bion cultures placed in the box gave bluish moving vapours and light-yellowish points and lines. But the experimentors were puzzled to note the same visual phenomena in the box after it had been ventilated; then in the box after the parts had been taken apart, aired, and the metal plates dipped in water to eliminate all traces of orgone energy; and finally, in a completely new box built for the purpose, which was kept away from the SAPA cultures.

These experiments served to emphasize the deduction that had first been made from the work with the electroscope, that the energy came from the sun and also was present in the atmosphere everywhere. It may be remembered that rubber gloves which had never been near any SAPA cultures, produced a reaction on the electroscope. This reaction could be eliminated by ventilating the gloves in the shade or by dipping them in water, and the reaction could then be restored either by placing them near the SAPA cultures or by sunning the gloves.

The next clue, in the discovery of the orgone, came from observations of the stars and sky made through an open wooden tube. Looking at a patch of dark sky, between stars, Dr. Reich noted the same flickering and flashing which he had observed in the specially-constructed box. A magnifying glass used as an eye-piece in the tube, magnified the rays. He came to the conclusion that "The energy in the box, in the absence of cultures, came from the atmosphere." The question of cosmic rays was considered, but cosmic rays have never been visible.

The Discovery of the Orgone, Volume II, contains a considerable amount of material on the visual observation of orgone energy, and depicts a device, the "orgonoscope", which facilitates those observations. It will suffice here to mention that through the effort to construct chambers for visualizing the orgone energy, it was found that those very chambers, built of metal inside walls and organic outside walls, served as devices to concentrate this new type of energy and make it available for experimentation and use.

First, we will give reports of a few of these experiments which prove that a definite form of energy is concentrated by these devices.

The observed fact that radiation from SAPA bions killed cancer cells, made it evident that this new type of energy might be very valuable and useful if it could be controlled at will, and if its characteristics and effects could be discovered.

5. ORGONE ACCUMULATORS

The boxes, which first started out as observation chambers, were later termed "orgone accumulators" for, in essence, such they turned out to be. Furthermore, it was found that a greater concentration of orgone energy could be attained inside those boxes, if the number of layers of walls was multiplied. The greater the number of walls, always alternating between metal and insulating material, the stronger was the accumulation of orgone energy inside the box. As the experiments progressed, both the size of the boxes and the number of layers were increased, until orgone accumulators the size of a room were built, and up to twenty layers were incorporated in the construction.

Experiments with these boxes and rooms produced a vast fund of new, vital and startling facts.

Orgone Accumulators and Temperature Effects

One of the evidences that odd and unusual things happen when these accumulators are used, is the peculiar set of temperature effects observed with these devices.

To quote again from *The Discovery of The Orgone*, Volume II: "The metal walls of our orgone accumulator are 'cold'. If we hold our palm or tongue at a distance of about 10 cm. (about 4 in.) from the wall, we feel, after some time, warmth and a prickling sensation. On the tongue, we receive a salty taste."

A thermometer inserted in the top of the accumulator registers a higher temperature than a thermometer placed in the room containing the accumulator. There is no source of heat in the accumulator itself, during the temperature tests. The temperature differences range from 0.2 to 1.8 degrees centigrade, and average 0.5 degrees centigrade. Difference in atmospheric humidity appears to account for the *variation* in the temperature difference, the lowest differential occurring on humid days, and the highest differential on dry days.

The theory, propounded to explain this temperature differential phenomenon, is that stoppage of the kinetic energy of particles results in a temperature rise (conversion of kinetic energy into heat)—the organic covering of an orgone accumulator takes up energy from the atmosphere and transmits it to the metal inside. The metal radiates the energy both outside and inside. That radiated to the outside is absorbed by the organic outer layer, while that radiated by the metal to the inside moves freely. *Thus, movement of energy toward the inside is free, while toward the outside it is being stopped.* It is believed that this stoppage of kinetic energy is what causes the temperature rise. This shows how these special structures act as orgone accumulators—orgone energy enters the accumulators but cannot leave them; therefore, the concentration of orgone inside the accumulators piles up, until inflow is neutralized by leaks through ventilation spaces. Boxes built only of insulating material show no temperature difference between inside and outside atmosphere; only boxes lined on the inside with metal show this difference in temperature. ... At any rate, the temperature difference obtained from such boxes is one proof that an energy phenomenon of a new and unique type is present.

ORGONE ACCUMULATORS AND MAGNETIC EFFECTS

If a magnetic needle is brought close to a multi-fold orgone accumulator, the north magnetic pole of the needle turns towards the centre of the lower edges of the accumulator. ... If we were dealing here with the ordinary magnetic effects, the magnetic needle would always turn towards the centre of the edges in the same way, no matter how the accumulator was turned. Repeated experiments showed that is not the case. When the accumulator is turned in various positions, whichever side is made the upper one is the side which then attracts the north magnetic pole. It would appear that the reaction depends upon the position of the orgone accumulator in relation to the field of the orgonotic atmosphere of the earth.

ORGONE ACCUMULATORS AND ELECTROSCOPIC EFFECTS

Charged electroscopes discharge more rapidly in strongly ionized air—that is, air which contains a greater quantity of electrons (negative electric units). It is assumed this is because the electrons help to form a path between the various parts of the electroscope, providing for a means of discharge. . . . Repeated tests with comparative electroscopic measurements inside and outside the orgone accumulators, under conditions otherwise identical, show that electroscopes discharge *slower* when placed inside an orgone accumulator. This indicates that orgone energy is different from ionization, since its effect on the electroscope is the opposite of that of ionized air. Electroscopes furnish another means of positive proof of the energy effect of the orgone, along with temperature and magnetic phenomena.

Therapeutic Effects of Orgone Accumulators

When a person sits inside an orgone accumulator, preferably a few inches distant from the inside walls, two groups of effects take place. One set is subjective—after a time, a feeling of warmth occurs, later a prickling sensation and, if treatments are repeated daily for a time, a feeling of invigoration occurs.

The objective effects are shown in the steady improvement of a wide variety of disease conditions for which the accumulators have been used by medical physicians for therapy. Effects particularly noticeable are:

    Improvement in the blood of the patient
    Rapid healing of wounds or injuries
    Rapid elimination of infections
    Shrinking of cancer growths
    Increase in energy

Use of Orgone Localized Applications

This is done by channelling orgone energy from a small box, about one cubic foot of inside space, into a metal funnel which is applied to the area of the body that is affected, when only a small area is involved. Rapid healing effects have been noted as a rule.

Why does Orgone Energy Have a Healing Effect?

For an understanding of this question, we turn to the pioneering studies in bacteriology performed by Dr. Reich. Our starting point is the bion, the living vesicles which he demonstrated could be produced from inanimate or so-called "dead" matter.

6. THE T-BACILLI

While observing the bions under a magnification of 3,000 ×, it was found that plain carbon particles remained black and did not take a biological stain, but the coal *bions* reacted positively—turned blue with gram stain. Not all of the prepared particles reacted; only those which had already undergone a certain degree of bionous development (thin membrane, increased fluid, and blue colour inside); the under-developed particles reacted neutrally, the same as the control preparation of carbon particles that had not been heated and cultured.

In observing the stained particles at 3,000 × magnification with oil immersion, it was found that most of the blue vesicles which previously showed a variety of shapes, now had assumed a spherical shape of about one micron in diameter. In addition, smaller *red* bodies were noted, of elongated shape, pointed at one end, and grouped around the larger, round blue vesicles. These red bodies, gram negative, ranged down to $\frac{1}{5}$ micron length, around the limit of microscopic visibility; they were not found in fresh preparations. The small, elongated red bodies were termed "T-bacilli."

Characteristics of T-Bacilli

It was found through extensive laboratory tests, that the T-bacilli originate from degeneration and putrid disintegration of living or non-living protein. They differ from cultures of rot bacteria, in that the T-bacilli agglutinate only after many months, whereas rot bacteria agglutinate in a few days. The T-bacilli have a deadly effect. Injected into mice in high doses, they were found to kill the animals within twenty-four hours.

Relation of T-bacilli to Diseased Conditions. Dr. Reich shows that any kind of cancer tissue, fresh or old, contains T-bacilli on microscopic examination. If boiled, the tissue disintegrates almost completely into T-bodies, with their characteristic red gram reaction.

Pre-cancerous tissue likewise shows T-bodies, though small.

Degenerating blood, when prepared as described by Dr. Reich in the book cited, likewise displays these T-bacilli.

In making bion preparations, one always obtains two parts of bions: the larger, blue bions which are now termed the PA bions, and the small black T-bacilli (they are black before staining; red after taking the gram stain).

When T-bacilli culture is added to a new preparation of earth, iron or coal bions, it will be seen under microscopic observation at 400 × in dark field or 2,000 × in ordinary light, that the T-bacilli that are near the blue bions show a restless activity; they turn around and around, then remain with trembling movements, in one and the same spot, and finally become immobile. As time goes on, more and more T-bacilli conglomerate around the blue bions; they agglutinate. The "dead" T-bacilli seem to attract and kill the living ones.

One of a series of experiments, reported in the book cited, concerns the injection of PA bions and T-bacilli in mice. In the first test, T-bacilli were injected in sufficient quantity to kill over one-third of the mice in one week, and to result in thirty more dying in fifteen months, with the remaining twenty-four ill during that period. . . . In the second test, PA bions were injected, then T-bacilli, in forty-five mice. None died within a week, nine died within fifteen months, the rest were healthy after fifteen months.

Other microscopic observations showed that the PA bions issue a radiation which destroys subtilis and proteus bacilli.

Implications of the Action of PA Bions upon Bacteria and T-bacilli

The PA bions represent fully developed, highly-charged orgone units. The T-bions, on the other hand, contain only very small amounts of orgone; they are weak orgonic systems, and develop either through primary insufficiency of free orgone within an energy vesicle, or degenerate from a higher form of energy vesicle through loss of part of the original orgone charge.

Now we begin to understand why the exposure of a living organism to the radiating energy of an orgone accumulator can have therapeutic effects. The PA bions represent the health principle; the T-bions the disease principle. T-bions cause disease when injected into healthy organisms. Diseased organisms contain too large a proportion of T-bions in their tissues. PA bions kill the T-bions, as proved by experiments microscopically observed. PA bions have a normal quota of orgone energy; T-bions are deficient in orgone energy.

By supplying additional orgone energy to the living organism, the process by which PA bions degenerate into T-bions is evidently stopped, and the PA bions are given a better chance to fulfil their function of killing the T-bions. Further light on these processes is furnished by the special orgone blood tests.

7. THE ORGONE AND HEMATOLOGY

Orgone and the Red Blood Cells

Bion research shows that the living red blood corpuscles, the erythrocytes, are orgone vesicles. Observed at a magnification of over 2,000 $\times$, they appear *blue* and show pulsation. Dead erythrocytes are black, immobile, and no longer pulsate. It is stated that the motility of the living erythrocytes can derive only from the internal (orgone) energy charge, not from external impulses. When the blue orgonotic colour disappears, the motility also is gone.

Quoting from *The Discovery of The Orgone*, Volume II:

"The biological vigor (functioning power) of a cell is determined neither by its structure nor by its chemical composition. The disintegration of structure and of chemical constitution are the *results* and not the causes of the biological disintegration, for the structure as well as the biochemical equilibrium of the cell are themselves expressions of its biological functioning capacity. The biological function itself has hitherto been a mystery. The orgonotic charge of the cell now makes it possible for us to determine the biological functioning capacity experimentally.

"The red blood corpuscles (erythrocytes) of two individuals may be the same structurally and chemically and yet differ sharply in their biological functioning. Microscopic examination may show exactly the same form; the number of the erythrocytes and their hemoglobin content may be normal and the same in both individuals.

The T-blood Test

"Now, let us expose samples of the blood to the same destructive influences. We autoclave a few drops of the blood in bouillon and Kcl solution for a half-hour at 120° C. and 15 lb. steam pressure. Microscopic examination may reveal widely different findings. The autoclaved blood of the one individual has disintegrated into large blue vesicles, while the blood of the other individual shows no blue vesicles but only T-bacilli. Gram stain shows the same differentiation: the one blood gives blue, gram-positive vesicles; the other red, gram-negative T-bacilli." Microphotographs illustrating these phenomena, appear in the book cited.

"The one blood sample points to a strong orgonotic charge of the erythrocytes. After autoclavation, this charge shows itself in the blue bions ("B" reaction). The other blood sample points to a weak or minimal orgone charge of the erythrocytes. After autoclavation, this lack of charge shows itself in the absence of blue bions and in the presence of T-bacilli; the result of degeneration of the erythrocytes ("T" reaction).

"The T reaction is typical of advanced cancer patients in whom the orgone of the blood has been used up in the struggle against the systemic disease (cancer biopathy) and the local tumour. The T reaction is often present *before* any symptoms of anaemia, and can betray the cancer process long before the development of a palpable or visible cancer tumour.

"Conversely, orgone-weak erythrocytes take up orgone avidly when it is introduced into the organism in the orgone accumulator. We find then that the autoclavation test reveals a disappearance of the T reaction and its replacement by the B reaction. That is, the erythrocytes become more resistant to autoclavation; they contain more orgone."

Charging of Erythrocytes with Atmospheric Orgone

"This can be checked experimentally. We mix a microscopic slide biologically (i.e., orgonotically); weak blood with rot bacteria or T-bacilli. Due to the weak charge of the blood, it does not kill or agglutinate the bacteria and the T-bacilli. If, now, we charge the organism orgonotically (the amount of the charge can be judged by the autoclavation test), we find that the blood has acquired a strong capacity to paralyze and kill the same pathogenic micro-organisms. The same phenomena can be observed with smaller protozoa that are not damaged by orgonotically weak blood but are paralyzed by strongly charged blood.

"The erythrocyte is a tiny orgonotic system which contains within its membrane a certain amount of orgone. If observed at 4,000 ×, the erythrocytes show a blue glimmer; their contents show a strong vibration; they expand and contract; in other words, they are not rigid as is commonly assumed. They carry the atmospheric orgone from the lungs to the various tissues. The connection between atmospheric oxygen and orgone can as yet only be guessed at ... The orgonotic charge of the erythrocytes is also expressed in their shape and structure. Poorly-charged cells are more or less shrunken and have a narrow margin of blue which is not intense. When the organism is charged, the erythrocytes fill out, the blue margin becomes intense and wide, sometimes extending over the whole cell. No pathogenic micro-organism can exist in the proximity of such strongly charged erythrocytes. ...

"That it is the orgone charge of the erythrocytes which exerts a killing effect on protozoa and bacteria is shown by the fact that in the process of killing the pathogenic micro-organisms, the erythrocytes gradually lose their blue colour, become black and sometimes disintegrate into T-bodies. The examination of cancer tissue in treated mice shows that the charged erythrocytes penetrate into the cancer tissue which, in their proximity, disintegrate into immobile T-bodies. But we no longer see any erythrocytes, only the T-bodies. The cancer tumour shows cavities which on microscopic examination (dark-field, 3–400 ×) are shown to be filled with T-bodies. ...

"Summarizing, we can say that the strongly orgone-charged erythrocytes act upon bacteria and small protozoa just like bions, e.g., of earth, iron or coal. Since they (erythrocytes) are formed in the bone marrow, we must assume that the bone marrow has the ability constantly to create new bions. The organization of energy vesicles is a basic characteristic of animal and plant tissue. These facts underlie the orgone therapy experiments in cancer: *by introducing orgone from the outside, we relieve the organism of the* burden of having to use up its own orgone in the fight against the disease. ..."

Additional notes:

The growth of protozoa from grass infusions is markedly *inhibited*, if the infusion is kept in an orgone accumulator from the beginning.

"Apparently, the orgone charges the grass tissue and prevents its disintegration into protozoa."

Fully developed protozoa are not killed in the orgone accumulator.

T-bacilli are not killed in the orgone accumulator, but the blood of cancer patients becomes free of T-bodies, if the patient has been intensely orgone-irradiated.

Those familiar with the biological and bacteriological sciences as taught in academic circles, will realize the revolutionary impact of the Reich methods and results. No doubt more light will be desired, to clarify the characteristics of the orgone energy manifestations. Further detail therefore follows.

8. INTERACTION OF BIONS

Radiation and Attraction between Two Orgonotic Systems

At first, earth bions and erythrocytes, both diluted to make observation of individual bions easier, move individually by themselves. "Gradually, however, they begin to group themselves, usually in such a manner that several erythrocytes collect around a large earth bion and come closer and closer, until they touch each other. Then, an intense radiation sets in where they touch each other. In places where the bodies are not touching each other directly, but are at a distance of about one-half to one micron, there develops a strongly radiating bridge between the earth bion and the erythrocyte which seems to connect the two. This bridge shows intense vibration and is alternately wider and narrower. Finally, the membrane between the two bodies is less sharply defined. If one only observes long enough, one can easily see how the erythrocytes begin to refract the light more strongly, their blue becomes more intense, they become bigger and tauter ahd show vivid pulsation. In this manner, one can *charge* erythrocytes with orgone just as one does in the body by orgone irradiation of the organism. If, in this experiment, one uses weak, deformed erythrocytes from cancer blood, their filling up and radiating is even more easily observable. Orgonotically weak erythrocytes exert little or no influence on bacilli and small protozoa. If they are charged with orgone, the effects appear. The erythrocytes "fill themselves" with the orgone from earth bions."

"Injection of sterile earth bions in cancer mice had the same effect as irradiation in the orgone accumulator; inhibition of tumourous growth, replacement of the tumour tissue by strongly radiated blood, and killing of T-bacilli. In the bion mixture, we can observe directly what takes place in the organism as a result of bion injections. This mode of orgonic application was used—before the atmospheric orgone was discovered."

Fusion of Bions

"Between earth bions and erythrocytes no fusion of substance takes place; only the formation of a radiating bridge. The same is true of iron bions, coal bions, and so on. However, coal bions and bions from autoclaved blood or any protein substance penetrate each other. This *fusion* will be shown to be of great importance for an understanding of the experimental production of tumours in mice by tar."

Dr. Reich and his staff succeeded in demonstrating fusion of earth bions and coal bions. "When the coal bions have been attracted by the heavier and thus less mobile earth bions and have formed the radiating bridge, we see the energy process ... At the points of the radiating bridges, the coal substance (in the form of bions) gradually begins to enter into the earth bions. It looks as if the earth bions were absorbing the coal bions. Finally, the smaller coal bions have penetrated completely into the body of the earth bions. They still can be distinguished clearly by their black membranes, while those of the earth bions are brownish. The whole complex, consisting of earth bions plus coal bions, looks brown and black. Gradually the black disappears; the membranes of the coal bions are dissolved. The earth bion becomes darker, its blue vesicles radiate more strongly. Finally, no coal substance is any longer visible."

"The fact should be mentioned that there is such a thing as a satiation of the orgone hunger in bions. If one adds only a few coal bions to an earth bion solution, one finds no coal bions left after a few days. But if one adds a large amount of coal bions, they do not all disappear.

"Different kinds of bions show different degrees of 'orgone hunger'. Sand bion cultures, for example, are 'greedy' towards coal bions; similarly, iron bions fuse readily with coal bions. Bions from boiled organic substances, such as muscle, are far less ready to take up the coal bions. These observations allow the conclusion that the less carbon a bion contains originally, the greater its tendency to take up carbon. The SAPA bions, derived from sand, originally contained no carbon, the iron bions only traces of carbon. The muscle bions, on the other hand, are composed of carbon compounds; thus their hunger for carbon is far less than that of the sand bions.

"In summary, the orgone energy vesicles show the basic functions of living substances fully developed: attraction, lumination, radiating bridge, fusion and penetration. These functions are specific characteristics of the orgone vesicles, for bions which have lost their charge also lose these functions. These functions, then, are not determined materially, but by *energy*. They are specific orgone functions and have nothing to do with magnetism or electricity."

9. PROTOZOA DEVELOPED FROM BIONS

Production of Protozoa

Dr. Reich denounces as unscientific and without foundation in fact, the theory that protozoa are produced spontaneously from "air germs". He performs various experiments to show that such a means of organization does not occur. In contrast, he reports in detail on the micro-biological techniques he has developed to demonstrate the organization of protozoa from grass infusions. In working with these infusions, he states that in the course of two or three days the grass disintegrates into bion vesicles, like any other substance that is made to swell. After another two or three days, the original cellular and striated structures can no longer be seen.

"... We focus our attention on the bions. We see how they conglomerate here and there into heaps and develop a membrane. We can follow every step in this development. Here and there we see some bion vesicles within a heap develop delicate rotating or vibratory movements. They assume an increasingly taut form and actually look like cysts. But they are not dried-up protozoa; they are forms in the process of developing from a bion heap. These bion heaps may be of various sizes and shapes. The tauter they are the more they assume spherical shape. They have filled with fluid—that is, they have become mechanically tense. The first phase of the function of tension and charge has taken place. ... The development of such a heap of vesicles into a pulsating protozoon takes one to two days. The protozoal primal vesicle (bion heap) remains immobile for hours—it gets tauter and tauter and more and more demarcated from its surroundings. Gradually, a movement of the energy vesicles within the bion heap begins to set in."

Then follows a description of four different types of observable movement:

Rolling—in which the energy vesicles within the bion heap roll rhythmically towards and away from each other.

Rotation—in some bion heaps, the total content of the vesicles start to rotate in one direction, and may continue the movement for hours. The rotation then becomes more intense, and finally one sees the whole heap rotating, including the membrane. "In this process it detaches itself from the surrounding grass tissue."

Confluence of the Energy Vesicles. "Not all of the bion heaps maintain the vesicular structure of their plasma. In many kinds of amebae, the boundaries between the individual vesicles disappear and the plasma forms a homogenuous mass with a bluish glimmer. In others, the vesicular structure remains until they are fully developed."

A variation of the *confluence of bions* can be observed in the development of many paramecia. "Here, the small bions do not flow together into one mass, but instead, groups of them form medium-sized vesicles within one larger body. These vesicles then show movements of rolling and rotating in relation to each other just as the bions from which they originated."

Pulsation, noted at a magnification of $3{,}000\times$, is in the form of very small movements of expansion and contraction in the bion heap. Forms in which the energy flow together are more likely to develop pulsation than those which maintain the vesicular structure of the plasma.

At the margins of the disintegrating grass, one finds every stage of development and form. As the transition from one stage of development to another is tiring to follow, due to the length of time involved, it can be shown more readily by periodic motion pictures which give in a few minutes the processes requiring wo to three days.

Differentiation between Protozoa

Up to the formation of the bionous primal vesicle, the development is observed to be the same in all forms of protozoa. After that formation, different forms of protozoa develop, and there is a correlation between the type of motion observed and the variety of protozoa that develops. The primal vesicles which tend to rotating movement and have a structure of large vesicles, usually develop into paramecia. The resting primal vesicles, with a flowing content, develop into amebae limax.

Nourishment of Newly-produced Protozoa with Bions

"The fully developed protozoa take up bions from the fluid by means of attraction. The attraction exerted by paramecia and colpedia on energy vesicles is enormous; it cannot be explained by the mechanical movement of cilia. For the vesicles in the fluid do not move past the body as one would expect from the action of cilia; rather, when they come within a certain distance, they are led towards the paramecium with great force. The observation is unequivocal. The org-protozoa contract, then expand and open the mouth widely—whereupon the bions in the fluid flow into it with great force. The mouth closes, the animal again contracts into spherical shape, and there is a rhythmic 'grinding' movement of the energy vesicles.

"The body of the protozoon shows a field of orgone energy which affects its surroundings. It affects bions, small bacteria and other small protozoa mostly in the sense of attracting or paralyzing them. Charged erythrocytes seem to be orgonotically stronger than paramecia or small amebae, for they are capable of reducing the mobility of these organisms. The protozoon consists, from the point of view of orgone physics, of a nucleus, a plasmatic periphery and an orgone energy field; it thus forms an 'orgonotic system'."

The protozoa developed from grass infusions, as described in foregoing paragraphs, exhibit the ability to procreate by division.

Production of Protozoa from Free Orgone Energy

Among the parade of wonders, issuing from the Wilhelm Reich research, one of the most startling is the achievement of the organization of protozoa from free orgone energy, in contrast to the methods already referred to, in which the protozoa were produced from matter that (as Reich expresses it) had already been "organized". Reich is particularly interested in this more radical method, as it provides further proof of the life-specific nature of the orgone energy.

It all started from a lesser discovery, that orgone energy possesses the property of lumination, and the intensity of fluorescence in fluids therefore provided an index of the extent of lumination present. The steps in this new way to fix the organization of protozoa follow, in condensed form from the book cited:

(*a*) Fluorometric intensity (and therefore the orgonotic potency) of fluid, which had contained earth bions, was much higher than that of ordinary water.

(*b*) Earth bion water of high fluorometric intensity that was kept frozen in sealed ampoules for several weeks was found to develop flakes which, after thawing, showed vigorous bions. These had sprung from water that had been carefully filtrated and by microscopic observation was free of particles.

(*c*) The water for this purpose has had screened garden soil boiled in it for an hour, or autoclaved for half an hour at 120° C. and 15 lb. pressure. The water is then filtrated from the boiled soil. This clear fluid is termed "bion water". Its fluorescence, measured by fluorophotometric equipment, averages forty-five times as great as the value obtained before boiling. The process imparts a yellow colour to the water. The process of boiling the earth in the water, liberates energy from the matter (earth) and transfers it to the water.

(*d*) To prevent the development of rot bacteria in the boiled bion water, it is autoclaved again, and then placed under refrigeration in sterilized containers. After a period under refrigeration, which may vary from days to weeks, flakes of whitish and brownish colours are found, of sizes ranging from one to five millimeters in length. Microscopic examination of these flakes shows both active bions, and strongly bionous heaps of orgone energy vesicles. More of them develop over a period of weeks.

(*e*) Conditions have been found under which these bions first enlarge to two or three microns in diameter, and then elongate to form bean-like particles. These "beans" can then develop into contractile protozoa which move rapidly in a jerky manner. The plasma of most of these protozoa has a granular or striated structure; in others it is smooth, without structure. One can obtain pure cultures of these protozoa by inoculating from the fluid *above* the flakes, without stirring up the flakes themselves. They increase from culture without difficulty.

Many tests, proving the organic and living nature of the bion flakes, are described in the book cited. The significance of the transfer of orgone energy from earth to water and its development into conventional living forms can scarcely be over-estimated. This is one of a series of developments which leads Dr. Reich to term orgone energy the primordial cosmic energy.

10.　THE CANCER PROCESS

I. Cell Findings

The belief that the cancer cell develops directly out of healthy tissue is erroneous. Long before the development of the first cancer cell in the organism, there are a series of pathological processes in the respective tissue and its immediate surroundings. These local processes, in turn, are induced by a general disease of the vital apparatus, termed the carcinomatous shrinking biopathy.

In this research, Reich uses different techniques from conventional cancer research:

(*a*) By examining healthy and cancer-suspect tissues in the living state, rather than in a dead state, fixed and stained.

(*b*) By using at least 2,000× magnification, which gives detail not obtainable with lesser magnifications.

(*c*) By making frequent observations of living blood, excreta, and skin cells of the organism under study.

Notes from Reich's Microscopic Observations

In healthy living tissue and in healthy blood, examined at 2,000×, are found exclusively such cells as are described in the literature. . . . In blood, excreta and tissues from cancer cases, formed cells and unformed shapes that are never seen in healthy animals or in tissues and sputum of human beings.

In particular, there are striated or vesicular structures with a strong blue glimmer which look neither like cells nor bacteria. Many of those that are formed, show elongated, club-like or caudate shape. There are also quickly-moving and pulsating ameba, even in lung tissue from cancer patients.

In sputum from lung cancer patients, in fact from excretions from any cancer tissue, there are the T-bodies in quantities, the T-bacilli—which can be cultivated from degenerating tissue or blood, or from putrescent protein, or from degeneration of bacteria. It has never been possible to cultivate T-bacilli from the air. . . . Blue, PA bions are also observed.

Stages in Cancer-cell Formation

Extensive research on normal and cancerous tissues, and excretions, show that the following is the sequence of cell development in cancer:

(*a*) The first phase of carcinomatous tissue degeneration is the loss of normal structure through the formation of vesicles.

(*b*) The vesicular disintegration of tissue results in two basic types of bions: the blue PA bions, and the black T-bacilli.

(*c*) The next stage is the formation of a membrane around the bion heaps— in some cases the bions dissolve into structureless or striated blue plasma.

(*d*) Then follows the formation of the club-like, caudate-form cells which are the true cells of fully-developed cancer, according to Dr. Reich and his staff. They have found these cells in all forms of cancer, whether the growth is in bone, glandular or muscle tissue. Observation shows that the caudate shape is assumed long before the cells gain motility.

(*e*) The fifth stage is the attainment of motility by the caudate-shaped cells. The movements cannot be seen with magnification of less than 3,000 to 4,000 ×. They are slow and jerky, from place to place.

The cancer cells move by way of rhythmic contractions or by flowing from place to place. Many cancer cells have a tail and move in the manner of fish, as recorded by Dr. Reich on movie film.

(*f*) The infiltration process. Vesicular disintegration softens the tissue, which makes the infiltrating growth of cancer cells possible. . . . The first vesicular disintegrated cell group develops into cancer cell tissue. This in turn damages the surrounding tissues and causes its vesicular disintegration. This tissue, in its state of vesicular disintegration, no longer offers any considerable resistance to the cancerous growth and itself increasingly develops into cancer tissue. It is a matter of mutual interaction between already-formed cancer tissue and surrounding healthy tissue.

(*g*) The final stage is the liquefaction of the plasm and the development of flowing ameboid protozoa. In humans, life often terminates before this stage is reached.

Comment

The entire cancer process, as disclosed by the Reich discoveries, is indicated as a process of self-disintegration and auto-infection of the organism. Disintegration of organized matter produces bions, as proven by numerous experiments often repeated by the Reich staff. Whether bions thus developed are the healthy, blue PA bions or the deadly, black T-bacilli, depend upon the orgone charge of the bions. . . . Bions under certain conditions develop into protozoa, as has also been proven by other Reich experiments referred to in this section and described in much more detail in the orgone literature. . . . The cancer cells are shown to be a particular variety of protozoa, having a significant shape always seen in any type of cancer, regardless of its location in the body, and these cells when fully developed show motility or movement when observed under adequate magnification. . . . Thus in metazoal (many-celled) organisms, cancer is the process by which individual tissues disintegrate into protozoa. The fatal accompaniment of this metamorphosis of tissues into protozoa is the typical cancerous process of putrefaction.

II. Blood Findings

The Reich research group has devised many types of blood tests which have the outstanding advantage of differentiating between normal, pre-cancerous, and cancerous blood. A condensed version of the principal factors observed through these tests follows:

(*a*) *Shape and Size of Erythrocytes*

At a magnification of 2,000× or over, healthy erythrocytes (red cells) are taut and oval in shape. In pre-cancerous blood, the erythrocytes are smaller, and often round, and in some of the cells a shrunken membrane can be seen. In blood from cancer cases, the shrunken membranes are very noticeable, and T-spikes have been formed (poikilocytosis).

(*b*) *Width of Orgonotic Margin*

Healthy erythrocytes have a wide, intensely glimmering blue orgone margin. Diseased (shrinking) erythrocytes show a narrow and pale orgone margin.

(*c*) *Pulsation*

Healthy erythrocytes show pulsation, at 2,000× or higher magnifications. In shrunken erythrocytes (the type found in pre-cancerous and cancerous blood) the pulsation is weak or altogether absent.

(*d*) *Resistance of Red Cells in Salt Solution*

Healthy red cells retain their normal size and shape up to half an hour or longer, after being placed in physiological salt solution. "Shrinking erythrocytes, or those with an increased tendency to shrink, often disintegrate within a few minutes or even seconds, show an irregular membrane and the so-called T-spikes." The presence of T-spikes points to an advanced degeneration.

(*e*) *Products of Disintegration of Blood Cells*

Healthy erythrocytes disintegrate slowly in salt solution, rapidly on autoclavation, into *blue bions* (the B-reaction). Erythrocytes from cancer cases disintegrate into T-bacilli (the T-reaction).

(f) Results from Blood Cultures

Healthy blood gives no cultures of bacteria in bouillon. Blood from cancer cases gives cultures of rot bacteria and T-bacilli. These organisms can also be found, states Dr. Reich, by direct microscopic examination at 2,000× or higher, of the blood of cancer patients. He has published numerous photographs to illustrate these points. "All these findings make the blood particularly suitable for an early recognition of cancerous processes. This is not surprising, in view of the fact that the blood is the 'sap of life' which connects all organs into one whole and nourishes them."

III. Orgone Aspects

The Reich findings show that the orgone energy is the life-energy of the cell. Cells richly supplied with orgone energy are healthy and vigorous. In animals and man, when the orgone energy is plentiful, the red blood cells have full membranes, wide orgonotic margins and deep blue colour of those margins, and do not readily disintegrate.

The local cancer tumour develops in tissues poorly charged with orgone energy. Conventional cancer research has shown chemical changes in cancer tissue, such as production of lactic acid and excess of carbon dioxide which point to a process of suffocation. Bion research adds the energy viewpoint to the chemical viewpoint, and shows that "energy stasis" leads to a bionous disintegration of the cell substance, and that the cancer cell develops secondarily from the resulting bions.

There is a long description in *The Discovery of The Orgone*, Part II, "The Cancer Biopathy" detailing just how the energy stasis leads to bionous disintegration of the cells.

It is a matter of curious interest that traditional scientific organizations and research failed to disclose the vitally important knowledge about the origin and development of cancer which has been disclosed by Reich research. Failure of conventional research to discover these facts stems from two causes: differences in techniques, and differences in basic concepts.

Techniques of Cancer Research

Traditional cancer research works almost exclusively with stained dead tissue sections. But "neither the blue bions from which cancer cells develop nor the small T-bacilli into which they disintegrate" can be seen in the stained tissue sections. The Reich research works with the living preparations. . . . There is a parallel here with the work of Royal Raymond Rife, who discovered many new things about bacteria and viruses by observations with his special microscopes which view those organisms in their live state. His method of illumination with a specific frequency of light, selected to harmonize with the life frequency of the organisms, makes it unnecessary to stain and kill the organism before observations are made.

With magnifications below 2,000×, most of the significant phenomena discovered by Reich cannot be seen. Yet traditional cancer research rarely works with magnifications exceeding 1,000×.

The development of the Reich techniques for analyzing blood and tissues by determining the time required for disintegration, and observing the products of such disintegration, discloses phenomena hitherto unknown.

Concepts—(summarized from Dr. Wilhelm Reich's comments)

The failure of orthodox science to realize that protozoa can be organized naturally from living or non-living material, completely blocked an approach to understanding the cancer cell.

The erroneous hypothesis of "air germs" diverted the searchers' attention. Reich has performed hundreds of experiments attempting to make cultures from the "air germs", and finds that the theory of the development of protozoa from air germs is fallacious.

Medicine and biology have a mechanistic, purely physical-chemical orientation. They look for causes in *individual* cells, *individual* organs, *individual* chemical substances, and *dead* tissues. Thus the *total* function, which determines every detail function, remains overlooked.

Orgone research, on the other hand, shows that the cancer cell is a product of a defence reaction of the organism. One should not make the error of assuming that the T-bacilli are the *cause* of cancer, even though they are the immediate culprits in the process of cell degeneration. T-bacilli can be cultivated from the blood and excretions of healthy people, though in much smaller quantities than from the blood and excretions of cancer cases—also, from healthy people, the blood and excretions have to be disintegrated for a far longer time before the T-bacilli make their appearance. The vital point is that "the disposition to cancer can be determined by the biological resistance of the blood and tissues to putrid disintegration. This biological resistance in turn, is determined by the orgone potency of the blood and tissues, in other words, by the orgonotic potency of the organism".

"Accordingly, to the extent to which any process diminishes the orgone content or the orgonotic function of the organism or of any of its organs, it will increase the disposition to cancer."

Orgone research has disclosed many relevant observations and experiments in this regard also.

IV. CANCER CELLS AS PRODUCTS OF A DEFENCE MECHANISM

It is well known that irritation can provoke the onset of cancer, as from old scar tissue or from tissues exposed to chronic abrasion, ulcerous tissue, and so on. The production of cancer from the effect of irritating substances such as chemicals from the coal tar group is also well known. ... Extensive orgone research has disclosed that the difference between tissues which give way to cancer upon exposure to irritation, and tissues which withstand irritation without producing cancer, lies in the fact that tissues in the former group degenerate much more readily into T-bacilli. Furthermore, it is the tissues that are weak in orgone energy that degenerate easily into T-bacilli—tissues that are orgonotically strong are healthy and for these tissues a scar or injury has no disastrous effect.

Let us examine the process that occurs when the orgone energy has become weak in the organism and irritation of one kind or another has initiated the cancer process. The general assumption is that "normal cells change into cancer cells." Exact study of the development of the cancer cells shows, instead, that the cancer cell is the *result* of the fact that the tissue fights against the effects of the T-bacilli. "The first step in the development of the cancer tumour is not the cancer cell, nor the disintegration of the tissue into blue bions, but the mass appearance of T-bacilli in the tissue or the blood.'

"When the T-bacilli begin to form and to amass anywhere in the body, the organism reacts with mild but chronic inflammation. Occasionally, the accumulation of white blood cells alone can get rid of the T-bacilli. In other cases, however, the autoinfection with T-bacilli is too strong or the orgonotic defence of the organism too weak. The question is, what happens then? How does the affected tissue react in this case?"

The answer was learned from numerous experiments in which T-bacilli were inoculated into culture mediums of organic protein. The presence of the T-bacilli resulted also in the formation of large quantities of blue bions, products of the disintegration of the organic protein. The same result was obtained repeatedly from the inoculation of T-bacilli into mice—by dissecting in series from the first day to the tenth week, one can follow the development of the cancer cells from the blue PA bions ...

A key discovery in this research was the fact that the T-BACILLI, WHICH ARE A PRODUCT OF A PUTRID DISINTEGRATION OF ORGANIC OR LIVING MATTER, STIMULATE IN OTHER ORGANIC OR LIVING MATTER THE FORMATION OF BLUE BIONS. This bion formation has the function of reacting against the T-bacilli (see page 68). That is, the blue bions are a defence reaction against the T-infection.

"If there were no more to the process than this local formation of PA bions, this local B-reaction, the T-bacilli would be of no further interest. In the blood of healthy individuals, one often sees the blood platelets, which are nothing but blue PA bions—surrounded by dead T-bacilli. Occasionally, one sees leucocytes filled with T-bacilli."

"Now, the weaker the orgone charge of the PA bions, the more of them must be formed in order to get rid of the T-bacilli that are present. But the blue PA bions develop into higher biological forms, protozoa (see page 69) and among them, into cancer cells.

The cancer cell is in reality a product of the many PA bions which were formed from blood or tissue cells, as a defence against the local autoinfection with T-bacilli."

V. The Fatal Process of Putrefaction

"The cancer tumour in itself is harmless unless it develops in vital organs (brain, liver, and so on). For this reason, people with small cancer tumours often continue their normal life, without feeling ill. Many elderly people have cancer tumours which cause no complaints and which are not discovered until the post-mortem examination. The typical cancer pains and the general weakness do not set in until the *total* organism is affected to a considerable degree. After that, the deterioration is rapid."

This latter state is characterized by transformation of the cancer substance into putrid matter, rapid development of rot bacteria, disintegration of the rot bacteria into T-bacilli, general T-bacilli intoxication, putrid bedsores, putrid odour, death.

"The actually fatal process is not the growth of cancer cells but the secondary T-disintegration. While the tissue damage is at first localized and the T-bacilli small in number, the disintegration of the cancer tumour results in a gigantic acceleration and in a general spreading of the putrefaction in the body, putrefaction of the blood, and T-bacilli intoxication of the body fluid system. In contrast to the phase of tumour formation, this second phase—that of the disintegration of the cancer tumours into putrid masses—lasts only a few weeks.

"This distinction is of eminent practical significance. Once the stage of secondary putrid disintegration in tumours, tissues and blood has been reached, the T-bacilli are formed in such enormous quantities that any therapeutic attempts are doomed to failure. In the first phase, however, in which cancer tissue is being *formed*, orgone therapy is highly effective."

VI. Therapy in Cancer

Experience in treating cancer patients with courses of daily irradiations in the orgone accumulator, shows that the charging of the body with orgone irradiation results in certain cases in steady shrinking of the cancerous tumours, gain in strength, gain in weight, and general systemic improvement. Treatment is continued over a period of months, and the basic improvement as shown by orgonomic blood tests is remarkable. However, at the time Dr. Wilhelm Reich wrote *The Cancer Biopathy* (1948), the main problem in treating cancer cases with orgone was not the destruction of the tumours, but rather the question of how to eliminate the products of disintegration from the body. In cases that were too severe or advanced when orgone treatment was inaugurated, there would be a period of improvement, often lasting many months, during which the tumours would be eliminated or greatly reduced in size, but the period of improvement would be followed by a period of decline leading to death from poisoning due to the large quantities of T-bacilli released through the disintegration of the cancerous tissue. Contrast here, the results obtained in advanced cases by the use of the Lakhovsky multi-wave oscillators, outlined on pages 251–254.

Dr. Reich places greater emphasis on the role that orgone therapy can play in early diagnosis and prevention of cancer, than on the treatment of advanced cases. He outlines how the disturbance in orgonotic functioning starts in the organism long before there are any external symptoms evident to conventional means of diagnosis, and is indicative of a general bodily or constitutional disorder. By the time a local tumour is evident, it is, from Dr. Reich's viewpoint, already too late to treat cancer under favourable conditions. He continues, "For the same reason, *local* therapy by surgical operation, X-ray or radium does not reach the cancer disease. The extirpation of a tumour may be ever so thorough, but the process of putrefaction is not touched by it. These facts are of extreme importance for a future prevention of cancer by orgone."

Regardless of whether there may be more effective methods of treatment of advanced cancer, it would seem that the Reich orgone research has opened a vast new field of knowledge regarding the nature and origin of cancer process, and could point the way towards its prevention through periodic orgonomic blood tests.

VII. The Cancer Biopathy

We turn now from a consideration of the physical changes involved in the *development* of cancer, to an outline of the factors preliminary to the cancer process. The preliminary developments which are found by Dr. Reich to set the stage for the later production of cancer, are termed by him "The Cancer Biopathy". It is here that principles of psychiatry come into play. Dr. Reich analyzes various cases at considerable length to illustrate proof of the theory; here we can summarize only briefly the main points.

He finds that the cancer cases he has analyzed all show a similar set of attitudes or viewpoints, marked by years of inhibitions or restrictive attitudes, and by resignation to an unhappy or restrictive environment.

The reactions of the organism to the inhibitions and restrictions consist of muscle and nerve spasms—(He presents evidence indicating that nerve tissue actually contracts of itself, apart from the contractions of muscle tissue).

The muscle and nerve spasms result in blocking the free flow of orgone energy to different parts of the body. He finds that cancer tumours develop first in areas that have long been spastic, or are fed by energy which first has to pass through areas which have become spastic.

In the treatment of cancer by orgone methods, Dr. Reich is interested not only in the physical charging of the organism with orgone energy, but also in re-education of the patient to the end that the free flow of orgone energy to all parts of the body may be permitted, for the first time in many years.

Interesting case histories of very advanced cancer are presented, showing that when progress has been made in these directions and the point is reached where the patient becomes aware of new currents of energy in the body—or rather, the re-establishment of normal currents, particularly those leading to genital excitation—an acute crisis is often induced. If the patient is unable to accept the psychiatric concepts involved, the reaction to the re-establishment of normal currents of energy within the body takes the form of severe anxiety, renewed spasms, and accompanying reactions which fight the flow of energy and can defeat the entire aims of therapy. Crises of this nature often mark the start of the period of final decline, following an earlier period of marked improvement under orgone therapy.

VIII. COMPARISON OF REICH AND RIFE FINDINGS IN CANCER

In comparing the respective research of these two scientists, with regard to cancer, it is interesting to note the similarities—also to find that the differences are complementary rather than conflicting in nature.

Both have achieved special progress by dealing with living tissue and organisms rather than with dead, stained tissue; both find that higher than normal magnification is essential; and both regard cancer as a constitutional process induced by changes in the internal environment, rather than as a local ailment that later spreads.

As for the differences, Reich's diagnosis of cancer starts when the blood shows an excess of T-bacilli and a deficiency in orgonotic charge. Rife's work on the cancer virus cycle may well serve to fill in the gap regarding initial processes which lead to the excess of T-bacilli observed by Reich. What one sees, in microscopic work, is determined by the level of magnification used. Reich, with his higher magnification and special techniques, sees far more than has been observed by conventional cancer research centres. Rife, working with still higher magnification than Reich, was able to observe and learn much of a virus-bacterial cycle significant to the initiation of cancer, a cycle which was not observable with the equipment used by Reich.

Reich, a psychiatrist, places great stress on psychiatric factors in restricting the free flow of orgone energy within the body, with resulting increased susceptibility to cancer. Nevertheless, Reich recognizes that chemical changes take place in the early stages of cancer, and Rife's work throws a great deal of new light on the nature of these changes and the effects produced, since he has discovered that a change in the chemical medium of as little as two parts per million is sufficient to change one micro-organism into another, and that the cancer virus is part of a five-element cycle which starts with a bacillus which is commonly found in every normal human body. Whether that common bacillus changes into another micro-organism which can lead eventually to the cancer virus, depends, according to Rife's work, upon the chemical environment in the body.

As it has long been known that emotions and thoughts have an effect upon the body chemistry, it is very possible that Reich's viewpoint is correct—regarding the effect of certain emotions as the initiating cause of cancer—and likewise that Rife's work provides the clues as to how the minute changes in body chemistry can lead to such resulting destructive effects. It would certainly seem that both of these types of research merit further attention and investigation.

11. ELECTRONIC DETECTION AND MECHANICAL REGISTRATION OF ORGONE ENERGY

In the search for instruments or equipment that would give positive proof electronically or mechanically of the existence of orgone energy, it was demonstrated time and again that this energy acts differently from any type of energy previously known. We have already mentioned that it does not affect an electroscope when applied to it directly, but has a very pronounced effect on an electroscope when applied *indirectly*, through use first to "charge" a quantity of insulating material.

It was found that tubes filled with argon gas would react to orgone energy by luminating when approached with a plastic rod charged with orgone, but the lumination did not occur in argon tubes when the experiment was tried too soon after bringing the argon tubes into the orgone laboratory—the tubes first had to stay in the orgone atmosphere for many weeks before they "acquired" the ability to react by lumination. The assumption drawn by Dr. Reich is that the tubes first had to "soak up" orgone energy, before they could act as detectors for that energy.

A similar type of experience occurred with Geiger-Muller counters. When first brought into the Orgone laboratories, they did not indicate the presence of any energy that would raise the background count. However, after weeks and months in the orgone atmosphere, they started to register very pronounced increases in background count—far higher than had been noted by users of this equipment when the equipment was exposed to X-rays or atomic radiation.

The increase in background count which was noted with the Geiger-Muller counter equipment, occurred only under certain significant conditions, in addition to the length of time required to "condition" the equipment to the orgone energy. The reaction occurred when the counter-tube was in its metal container, or else in a special cylindrical orgone-energy accumulator built to hold the tube. Time and again the reaction ceased when the tube was removed from its container.

The reaction of sharply increased background count was also obtained when the counter-tube was brought within two feet of the walls of a regular orgone accumulator, or when it was brought close to Dr Reich's body, which was kept charged with orgone energy.

The extent of the increase in background count varied in accordance with factors which had previously been observed to affect the strength of orgone energy, such as humidity in the atmosphere, presence of fresh air, and the use of water. Humidity, fresh air, and water, all weaken the orgone effect. There also appeared to be subtle differences in the construction of the counter-tubes which made some of them react more strongly to orgone energy.

Special vacuum tubes were constructed, having two large aluminium plates facing each other, and a tungsten wire running through the centre of the tube, between the plates. These tubes, after being "conditioned" by placing them in the orgone room for a prolonged period of time, showed very strong lumination from both orgone energy fields and from an electric charge placed across the aluminium plates.

Nevertheless, there still appear to be variables which have not been discovered and which therefore cannot yet be brought under control, with regard to these "vacor" tubes, as Dr. Reich terms them. For example, all the Vacor tubes constructed before December 1947, reacted with the lumination phenomena as briefly mentioned here, and which has been much more fully described in the "Orgone Energy Bulletins". Then, Christmas, 1947, brought a big snowstorm. Tubes constructed during the year following that snowstorm did not show any lumination, although tubes constructed *before* the snowstorm continued to luminate satisfactorily. Then on April 7, 1949, one of the "dead" Vacor tubes luminated for the first time. It is puzzles like these which show how much remains to be learned about orgone energy.

However, a great deal of additional information about the nature of orgone energy has been obtained through various experiments performed with the Vacor tubes, as described in detail in the "Orgone Energy Bulletin", November 4, Volume 3, October, 1951.

Again we wish to state that there are many more aspects to the Orgone pictures than we can even outline here.

12. SOME PRINCIPLES OF ORGONE PHYSICS

Discoveries in the realm of orgone energy necessitate revision of some of the traditional theories of physics.

1. The indications are that orgone energy is everywhere, throughout all space, and that it pervades even the so-called "vacuum". It would seem that orgone is, in reality, the cosmic ether for which scientists searched so diligently several decades ago. Why did they fail to find it? Was not the existence of an "ether" supposed to have been disproven by the negative results of the Michelson-Morley experiments? On the contrary, Dr. Reich shows that those experiments were based on assumptions which orgone research have proven to be incorrect; therefore the results of the experiments were misinterpreted.

To be specific, the Michelson-Morley experiments were based on the beliefs:

(*a*) That ether is at rest.

(*b*) That light is transmitted in space from one location to another.

If these beliefs had been correct, then the existence of ether would have resulted in an observable phase difference between light beams sent in the direction of the supposed ether "drag" and the light beams sent perpendicular to it. This was the phenomenon which the Michelson-Morley experiments attempted to observe.

In contrast, Dr. Reich and his staff have shown that the ether envelope surrounding the earth is not stationary, but normally moves from west to east, more rapidly than the rotation speed of the earth.

They have further demonstrated that the concept of "light" is composed of two factors—"excitation" and "lumination"—and that "light" as such does not travel through space, only the "excitation" travels, and when it reaches an area where local conditions make possible an orgone lumination, then the visible phenomenon occurs which we have learned to call "light".

Thus, both of the main assumptions of the Michelson-Morley experiments have been proven incorrect, since ether is not stationary and light as such is not transmitted; therefore, their experiments were not appropriate tools for detecting the presence of an ether.

Dr. Reich's term for the ether is the "orgone energy ocean". This refers to the lowest level of orgone energy, which pertains everywhere except in the locations where greater concentrations of orgone exist because of special atmospheric conditions or the presence of living organisms, or concentrations artificially created such as in orgone accumulators or orgone-energy rooms.

2. One of the most interesting characteristics of orgone energy is its basic tendency to flow from a weaker orgonotic system to a stronger one. This is in direct contrast to mechanical and electrical energies which always run downhill so to speak—they flow from the stronger to the weaker, from the higher potential to the lower potential, whereas orgone energy, in contrast to previously known laws of physics, runs uphill.

Dr. Reich presents an interesting case for the viewpoint that this contrary action of orgone is necessary to support life, for he states that otherwise the living organism, which must maintain a higher energy level than that of its environment if it is to survive, would in fact promptly lose the energy level difference between itself and its environment! In other words, what we have here is a *cycle*—physics as previously known had discovered only half of the cycle—the means by which energy is transferred from a higher level to a lower, from a stronger system to a weaker—but the very existence of a higher level or a stronger system implied some means by which energy could flow from the lower or weaker up to the higher or stronger, and orgone provides the hitherto missing half of the cycle.

The entire process is graphically illustrated in the following diagram from the October 1949, "Orgone Energy Bulletin", page 147:

It seems to us that this provides a most important link between the physics of inanimate matter and the physics of living organisms. In fact without the orgone

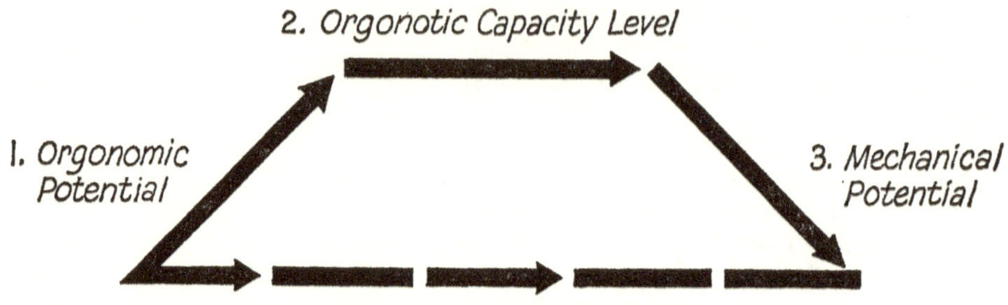

Diagram of the *Orgone Energy Metabolism* in living bodies.

theories, it is virtually impossible to integrate the energy aspects of living and non-living structures.

3. The orgone-mechanical energy cycle, as mentioned above, requires the revision of some of the traditional principles of physics, since it invalidates the Second Law of Thermodynamics, and some aspects of the law of Entropy. The law of conservation of energy needs to be restated. Reconciliation of this principle with the newly-discovered orgone phenomena, can be made if we assume that—

> "While some orgone units are forming in the orgone ocean by concentration, others terminate their single existence by energy dissipation into the orgone ocean. Thus the energy lost by discharge of 'deaths' of a number of orgone units would be picked up again to be concentrated in other units."

4. Observation of orgone energy movements show them to be made up of two different types of motion:

Discontinuous pulses, which move slowly from west to east.

Continuous waves, which move more rapidly in the same direction.

The pulses are superimposed upon the waves, like mountain peaks upon a mountain range.

The pulses and waves remind one of two chief characteristics that have been discovered of light—quanta (discontinuous pulses), and wave motion of a definite frequency.

5. An aspect of the orgone, which sharply differentiates it from other forms of energy, is its faculty for modification and change in characteristics in accordance with environmental factors. That is, it is sensitive to various types of influences, and changes its form of manifestation accordingly. This is something quite new: light, heat, radio waves, and similar energies can be absorbed by certain substances, but the energies themselves do not undergo changes in character except through absorption or through direct manipulation by means of specific equipment designed to accomplish a particular purpose. The orgone,

on the other hand, becomes excited or irritated when in the presence of metallic substances, and the reaction is even more pronounced when living substances are present in a metal-enclosed darkroom, or if there is electro-magnetic discharge from an induction-coil system placed in such a room.

The excitation or irritation of orgone is evidenced by its change in colour, in form, and in rapidity of motion, as observed in a darkroom. When not excited, it is bluish-grey in colour, fog-like in form. Excitation changes the colour first to blue-violet, later to yellowish-white—and changes the form to luminating dots, thence to rapidly darting rays.

The faculty of being sensitive to the environment, or being irritated by environmental factors to the point where it changes its colour, form, and speed, are characteristics which partake of the nature of living substance. And, in fact, the hypothesis is—that orgone is the true life energy, and the difference between a living and a non-living substance is primarily a difference of orgone metabolism.

HOMOEOPATHY

HOMOEOPATHY

GENERAL INTRODUCTION

THE Homoeopathic physicians have made possibly the most intensive of all studies of the nature of Man and the expression of that nature in health and disease. Much can be learned from an examination of both the older, classical form of homoeopathy and the newer developments in the use of certain classes of homoeopathic remedies.

All homoeopathic methods have in common the use of remedies prepared in homoeopathic form, so this will be our first topic for consideration.

1. *Preparation of Homoeopathic Remedies*

The substance to be prepared, whether of animal, vegetable or mineral nature, is ground into powdered form. One part of this powder is mixed with nine parts of an inert substance such as sugar or milk, and the mixture is vigorously ground together for a prolonged period of time. This makes what is termed a 1x preparation. Then one part of this mixture is mixed with nine parts of the inert substance and likewise ground together for a similar period of time—the result is termed a 2x preparation, and contains a 1/100 proportion of the original substance. The dilution process is repeated successively—thus the third round results in a 3x preparation, containing 1/1,000 part of the original substance—the sixth round results in a 6x preparation, which contains one-millionth part of the original preparation, and a 12x preparation contains one-trillionth part of the original preparation. This process is called "trituration".

Another method of homoeopathic preparation is to make a solution of the substance, using water or alcohol, and prepare successive dilutions on the same decimal scale as was mentioned above, giving to each dilution a particular number of knocks, or "succussions" as they are termed.

2. *Effect of Extreme Trituration or Dilution*

From a chemical standpoint, the quantity of therapeutic substance in a homoeopathic dose of 6x or more, is negligible. Yet, surprisingly enough, the greater the dilution of the substance, the greater is the effect of the remedy upon the body! In fact, the more times the substance is triturated or diluted, the higher is its potency, and from the very high potencies, which in one-grain doses contain according to mathematical physics not even a single molecule of the original substance, the physical reaction can be so pronounced in the body as to be very unpleasant if misused through dosage repeated too often or wrongly prescribed.

3. *Scientific Principles Underlying the Homoeopathic method of Preparation*

At first glance it might appear that the homoeopathic principle—that the smaller the amount of therapeutic substance remaining in a dose, the greater is its effect—is directly contrary to established laws of cause and effect. However, closer investigation discloses a number of scientific principles which clarify and support the results obtained from the homoeopathic method of preparation.

It is part of homoeopathic theory that the repeated application of mechanical energy to the therapeutic substance, applied in the form of grinding of triturations or succussions of liquids, results in spreading the molecules of the original therapeutic substance farther and farther apart, thus altering its fundamental nature and *releasing essential energy* (energy of its essence) and making that energy available to the human body

That such an effect (the alteration of the molecular structure of a substance) actually occurs through the repeated application of mechanical energy, has been proven by ingenious experiments in chemistry. For example, the French chemist Berthelot, working in laboratory in the College de France in 1856, demonstrated the production of alcohol by the use of the succussion process, with chemicals which when brought together without succussions would produce no alcohol at all. Specifically, he found that 900 grams of sulphuric acid would absorb 30 litres of bicarbonated hydrogen in the presence of a few kilograms of mercury, *if* subjected to 53,000 successions After this had been accomplished, the addition of water and the process of distillation resulted in the production of alcohol. He stated that the 53,000 succussions divide the molecules of the mercury and separate them more widely from each other. Then these molecules in their turn divide and separate the bicarbonated hydrogen and of the sulphuric acid, reducing the two latter to the *species of radiant state*, and permit the sulphuric acid to absorb the bicarbonated hydrogen.

Physics teaches that matter can exist in three different states—solid, liquid, and gaseous; it neglects the fact that some of our greatest physicists have taught that matter exists in four states—solid, liquid, gaseous and radiant. According to the eminent physicists Michael Faraday and Sir William Crookes, when matter is in the solid state its constituent molecules touch and are adherent among themselves. In the liquid state the molecules of matter still touch, but are not adherent. In the gaseous state, molecules of matter are not adherent and are more widely separated from all others. In the radiant state of matter the molecules are still more widely separated than when they were in the gaseous state. According to Crookes, radiant matter constitutes in reality the limit where matter and force seem to shade off into each other

Modern science has now demonstrated the ultimate in the extraction of large amounts of energy from small amounts of matter, through the release of atomic energy—but atomic energy is only a further step in the logical sequence of the development of the different states of matter; homoeopathy deals with the conversion of matter into energy in a gentle, constructive fashion whereas atomic physics deals with the conversion of matter into energy in a violent, explosive manner.

3. *Types of homoeopathy*

In this section we shall consider in turn:

    I. Classical Homoeopathy.

    II. Twelve Tissue Salts (Schussler's Method).

    III. Dr. Littlefield's Method of "Vitalizing" the Cell Salts.

    IV. Homoeopathic Bio-Chemistry.

I. CLASSICAL HOMOEOPATHY

1. INTRODUCTION

Classical homoeopathy involves:

    1. A concept of the nature of disease.

    2. A concept of the nature and purpose of adequate treatment.

    3. The special manner of preparing remedies, already described.

    4. A system of principles for the selection of remedies.

Homoeopathic concepts differ radically from allopathic concepts. (The "regular" medical physicians are termed "Allopaths".) Following are three important respects in which these concepts differ:

2. PALLIATION AND SUPPRESSION *v.* CURE

Allopaths often give drugs to relieve or palliate symptoms. Homoeopaths are directly opposed to the administration of drugs for relief or palliation, stating that such administration frequently drives disease deeper into the body, and leads to subsequent illnesses which become more and more serious as the suppression is repeated. Homoeopaths believe that the best "relief" comes from clearing up the *cause* of symptoms.

3. SPECIFIC REMEDIES *v.* REMEDIES SELECTED BY THE LAW OF SIMILARS

Allopaths select remedies for their effectiveness in suppressing a symptom. Homoeopaths work in the opposite direction—to clear the system of the disorder which produces a symptom. They do this through the use of a principle termed "the law of similars". For example, quinine given to healthy persons, produces symptoms similar to those of malaria, the very disease which it cures when given to those who have contracted the disease.

For every symptom, there is a drug which if administered in homoeopathic form will produce the same symptom if given to individuals in good health. It is part of classical homoeopathic theory that the energy from such a drug, made available to the patient through administration of the substance in homoeopathic form, is the only true cure for the symptom. However, in actual practice, the "totality of symptoms" must be used, as will be described later.

4. Gross or Physiological Doses v. the Minimal (Homoeopathic) Dose

When a drug is administered in gross dosage (in contrast to diluted or homoeopathic dosage), three types of effects occur:

(*a*) The so-called physiological effects. It is to obtain one or more of these effects that allopathic physicians administer a drug. For example, a physiological dose of atropine or belladonna, often given as an anti-spasmodic, also produces dilation of the pupils of the eyes, dryness of the mucus membranes, and flushing of the skin.

(*b*) Toxic effects, since the physiological dose of many drugs is not far below the maximum dose that can be given with safety.

(*c*) Other more subtle effects, on the functioning of the human organism and the feelings of the patient. These are of no interest to the allopath but are of prime interest to the homoeopath.

Considering these in turn, homoeopaths point out that allopaths, in order to obtain a specific physiological effect (in the case of atropine mentioned, the effect of relaxation of spasms), administer doses which produce other physiological effects that are uncomfortable, unnatural and often harmful to the body, as for example the excessive drying of the mucus membranes.

Under the heading of toxic effects, the use of gross dosage often results not only in immediate toxic symptoms, but also in clogging the body with drug residues. These residues can remain in the system and poison one or more of the internal organs, initiating degenerative processes leading to chronic disease later in life. The chronic disease can be much more serious than the acute ailment or symptom for which the drug was prescribed.

When a substance is administered in extremely diluted or homoeopathic form, there are none of the type of gross physiological effects which occur with allopathic dosage; there are no toxic effects, as the amount of therapeutic substance contained in a homeopathic dose is far below the minimum required to cause a toxic reaction, and there are likewise no residual quantities left in the system to cause trouble. An example of the damage done by residual effects is the case of sulfa drugs, which in the hands of allopaths caused crystallization of the kidneys and resulted in severe kidney damage to many, many persons through allopathic administration until sad experience forced the allopaths to reduce their administration of the sulfa substances and to develop less harmful varieties of the drug. Less immediate but more insidious poisonous effects are a constant concomitant of allopathic dosage of many different substances, according to homoeopaths.

With homoeopathic dosage, the effects mentioned in category #c—the more subtle effects which often are masked by the other effects of the massive allopathic dosage, are brought to the forefront, and these are the effects which are desired by the homeopaths. For some substances, they are often quite different from the physiological effects of a gross dose. The homeopaths have made a most detailed study of the effects produced by various substances when administered in minimal or homoeopathic dosage, and it is these effects which are taken into consideration when using the "law of similars".

5. DIFFERENTIATION OF SYMPTOMS

Matching the remedy to the symptom, in homoeopathic prescribing, is a much more detailed matter than might first be imagined. For homoeopathic physicians find that minute differences between symptoms are of the greatest importance since they indicate the need for totally different remedies. Homoeopaths therefore question their patients in great detail, and pay attention to factors which are usually ignored or considered unimportant by other schools of healing. For example, in one text we noted thirty-eight different listings of coughs each with its own indication of a remedy or remedies. If the cough is worse in the evening, one remedy is indicated—if worse in the morning, a different remedy is indicated. If it is better sitting up, one remedy is indicated—if better when lying down, still another is the more appropriate remedy. Still other listings go into detail regarding the nature of the cough, its location in the body, whether chronic or acute, whether dry, barking, hard or hoarse, whether paroxysmal or periodical, and so on through many other distinctions, each with its own indicated remedy or remedies.

Similar lists have been compiled in homoeopathic texts for every type of symptom known—for example under headaches, one book which we have noted contains over 130 different distinctions and differentiations. Often a symptom of a patient will coincide with several different listings. It may have several different characteristics, as for example in a case of headache might be considered the time of day when it is worse, the factors that aggravate it, the location of the ache in the head, the type or character of ache, etc. If the same remedy is included in the listings for two or more of the indications, the probability of that remedy being the correct one is increased. Under many of the symptom listings, there is a choice of remedies; sometimes a rather substantial list is given from which to choose. How is the choice made? Here we come to another of the principles of homoeopathic prescribing—namely, the "Totality of Symptoms".

A patient ordinarily has more than one symptom—to return to the example of headache, that may be the only complaint that a patient states, but the homoeopathic physician may also notice, for example, an abnormal colour of the face, pimples or other eruption on the skin, and questioning may disclose symptoms of intestinal irregularity, digestive disturbance of some nature, and possibly several other matters in which there is deviation from normal but which did not trouble the patient enough for him to mention anything about them. Now, for *each* of these secondary symptoms, the homoeopathic texts again have numerous differentiating listings or distinctions, and the careful prescriber will look up each separate secondary symptom in turn, find the exact listing or listings that correspond to the manner in which that symptom operates in the particular case under consideration, and note the list of remedies indicated for each of these listings. From all the lists thus selected, the remedy that is indicated the greatest number of times receives primary consideration in the selection, though other considerations are also involved. The point here is that the symptom of which the patient complains, in this case the headache, is only one factor in the total picture—and the homoeopath gets his best results by *treating the patient rather than the symptom*. Even with the multiplicity of listings under headaches, the remedy that will clear up the cause of a particular type of headache in one person will not do so in another having exactly the same type of headache, and the difference lies in the *other* bodily difficulties which are likely to be different in each case. Remember that the homoeopath is not interested in simply palliating or relieving the pain of a headache—he is interested in the cause which is back of the patient's tendency to have headaches at all!

6. The Three-fold Nature of Man

The above may be sufficient to indicate that the successful selection of remedies in homoeopathic practice is a very complex and difficult matter, but we have by no means covered all the aspects that are taken into consideration. We have referred only to the physical symptoms thus far, but now let us listen to Dr. Julia M. Green on Homoeopathic Philosophy:

"Man is a triune being. He possesses a soul and a mind as well as a body. His body is his house in which he himself, that is, his spirit and his mind, lives. When this triune being is all in order, the man is healthy. When this being is in disorder, the man is sick. It is far better to talk of curing an individual than to talk of curing a disease. The man himself is sick before his tissues are sick or changed, hence, to treat tissue change is superficial and is begging the question. Therefore, who is sick? The man himself who lives in his body. What is sick? The man himself, that is, his soul or spirit, his mind, his body; generally all three. Who is healthy? A man without disorder in all three parts of him; a man in order spiritually, mentally, physically.

"How can the inner man express his sickness? He turns his house into disorder; he has no other method of expressing his sickness. So, bodily expressions of sickness are expressed in signs, symbols or symptoms and through these we are continually seeking the sick *man*. These symptoms can be classified as to whether they express the spirit of the sick man, or his mind, or his body.—We say man sees, hears, feels, tastes, etc. but these are symbols through which he expresses himself, symbols of thinking, acting, living. Man thinks, wills, loves, understands; the dead body cannot. He uses his living body to manifest himself.

Therefore, it is the will and the understanding which constitute man himself. If they are in order, the man is healthy—if in disorder, he is sick."

From the above, the reader may already have guessed that the apparently over-burdened homoeopathic physician must consider not only the physical symptoms, but also the mental, emotional and spiritual aspects—and there are actually long lists of symptoms under these categories, each differentiated in accordance with minute detail, and with their own groups of remedies. The remedy to be selected must be that which not only comes closest to meeting the totality of symptoms on the physical plane, but also on the higher planes of mind and spirit as well.

Resuming our quotations from Dr. Julia Green, as reprinted from the Journal of the American Institute of Homoeopathy, November and December, 1941:

"We all know that illness is apt to change the patient's temperament. A steady cheerful person becomes irritable and unreasonable; or a patient endeavours to keep in usual emotional balance with the result that reactions take place physically as evidenced by disordered nerves, an upset digestion, disorganized habits, unusual behaviour of heart and blood vessels, etc. Disorder in the man himself necessarily reflects in all three parts of him. Until one's observation of these things is aroused, it is easy to say that a slight surface disturbance cannot possibly affect the inner man.

"Such effect can hardly be noticeable and easily thrown off if the inner man is healthy, that is, free from disorder, but even a tiny thing can produce alarming consequences if all is not harmonious within.

"To express such harmony, and the lack of it, we talk of susceptibility to disorder. When an epidemic comes along, some people come down with the epidemic disorder and others do not. It is all a question of susceptibility. If a man's resistance is strong, that is, if he is in harmonious order within, he does not 'take it' as we say. To express it another way, if his vitality is good, or if the inner man is busy dealing with another kind of disorder closer to his inner defences, he does not have the acute attack. The objection will be raised here that oftentimes it is the strongest and healthiest who succumb. Apparently true, until the whole state of the man himself is examined carefully.

"Every human has a soul or spirit, a mind or intellect, and a body. The spirit is the highest, most vital part of man; the controlling part. If there is a spiritual disturbance (emotional disturbance) this is able to disturb both the mind and the body and if strong enough, can demoralize them. The mind (intellect, reasoning power, memory) can seem to be the starting point of disorder and throw out of balance both the spirit and the body, but real healing comes to the spirit first. Likewise physical disorder can affect the spiritual and mental, but these start it and true healing will come from the highest or inner-most.

"The inner man who lives in his body as his house, is sick before his mind and body are sick, and the inner man must be healed first.

"In these days sick people go from one doctor to another, from one specialist to another, have 'general examinations' and are often then told that nothing is wrong, no disease is found. They are entirely unsatisfied because these people suffer much, are very miserable; they *know* they are sick. The homoeopathic physician treats such sufferers successfully because he knows that man is sick before his organs are sick, before a disease can be named.

"If disease in organs and tissues is all there is of disease, why cannot people be made well by removing diseased parts? We all know human wrecks who have had such surgery and still suffer.

"So it is the individual, the changes in the person, the patient, which have to be considered in order to accomplish a real cure. Homoeopathy is intensely individualistic." (End of quotation from Dr. Julia M. Green.)

From various of the foregoing passages, it will be seen that psychological and psychiatric principles play a part in homoeopathic analysis of cases, and that it is part of classical homoeopathic theory that the correct remedy, given orally, can accomplish what psychologists and psychiatrists try to do by verbal and other analytic treatment techniques. More on this subject will appear later in this section.

We are still on the subject of the factors taken into consideration in homoeopathic prescribing. In addition to the selection of remedies according to symptoms, the exact differentiation of symptoms, the use of all physical symptoms in addition to the one complained of, and the consideration of mental, emotional and spiritual factors in selecting a remedy, there are two other factors considered of vital importance by homoeopathic prescribers. These are, the environment of the patient, and his or her inheritance. The consideration of inheritance is particularly interesting as it involves an extension of the concept of suppression. Homoeopaths find that when one or both parents indulge in repeated suppression of ailments (and this is done by practically all who are under standard medical care) the disease tendencies are driven inward and are expressed not only by later illnesses but in inherited tendencies in children born to parents after such suppression has taken place. Many times, homoeopathic physicians find they cannot cure a person of his ailment until they first give him a remedy which one of the parents should have had long before, to clear up a condition which instead was suppressed.

7. Details of Homoeopathic Practice

The diseases treated, include the entire list of internal ailments. A true homoeopathic physician relies exclusively on homoeopathic remedies except where there are broken bones or severed arteries from accidents, which require mechanical intervention, or extreme emergencies such as danger of death from electric shock or from taking poison. Homoeopathy can often cure ailments for which allopaths resort to surgery.

To practice classical homoeopathy, a physician must first take the entire medical (allopathic) training, and then take post-graduate courses in homoeopathy as a speciality.

It is part of the theory of classical homoeopathy, that the only correct method of dosage is the use of the "single remedy". That is, only one remedy is to be administered at one time, and it is given full chance to show its effect before another remedy is chosen. In the course of clearing up a chronic condition, several different remedies may be needed, but they are administered in series over a period of time, and never consecutively. According to homoeopathic theory, the use of more than one remedy produces a conflict of vibrations which can obscure the picture rather than clarify it.

When prescribing homoeopathic remedies, the so-called low potency remedies (3x, 6x, and 12x triturations and dilutions) are directed to be taken daily or oftener, while the higher potency remedies are given at intervals of days or weeks. In some of the very high potencies, the effect of a single dose lasts for months, and the homoeopathic physician places great stress on letting all of the reactions from a dose occur, no matter how long a time is required, before giving another dose.

There are about 500 "proven" remedies commonly used in homoeopathic practice, and about 1,500 additional remedies that are less commonly used but which can be drawn upon when necessary for rare or difficult conditions.

8. The "Proving" of Remedies

The process of determining what symptoms a homoeopathic remedy will induce in a person of good health, is termed "proving". This process has been essential in building up the homoeopathic materia medica, or group of remedies suitable for homoeopathic use. It should be noted that homoeopathic remedies are always proven on *people*, not on animals as is the case with whole drugs used by allopaths; the reason for this practice in homoeopathy is that people are different from animals and do not react in the same manner as animals. It is the viewpoint of homoeopaths that the allopathic practice of giving drugs to people on the basis of apparently satisfactory results in tests on animals, has resulted in driving a great deal of disease "underground", only to cause more trouble later.

Since the process of proving is of interest, we are giving a brief description, again quoting from the homoeopathic physician Dr. Julia M. Green in her talks to laymen which were reprinted in the Journal of the American Institute for Homoeopathy, November, and December, 1941:

"In order to discover how drugs can express their power to help in the curative process, they have been given to healthy people to make them sick, to express their sick-making power; those well people who take drugs to make them sick are provers of those drugs. (The term drug is here used as any substance proven homoeopathically, whether animal, vegetable or mineral in origin.) The resulting symptom pictures are provings of those drugs.—

"The physician director of the proving gives the drugs in repeated doses to each member of a group of provers. In the group are men, women and children, many ages and many temperaments. They are as nearly healthy people as can be found. They do not know what they are taking and they do not communicate with each other during the proving. Nearly all of the group receive the drug to be proved. A very few receive blanks to be used as controls. Each prover is instructed to observe carefully every bit of change from normal and to report such changes regularly to the director. He is told nothing is too trivial to report.

"The director writes down what the prover reports in the prover's own language. Finally he makes lists of all symptoms reported and arranges them according to their spiritual, mental and physical aspects. He notes those symptoms felt most strongly by the largest number of provers and gives them the highest rank in proving. Generally there are three or four grades coming through. He gives the fine, sensitive provers a chance to bring out the finer nervous symptoms not felt by the more phlegmatic provers.

"When all symptoms are reported and no more are experienced, the proving is over. This may take many weeks. The director then arranges the symptoms according to the part of the body affected, e.g., the mind, head, eyes, ears, nose, face, throat, stomach, abdomen, pelvic region, chest, back, extremities, etc. There are also groups about sleep, dreams, nerves, weather, fever, sensations, etc., from the innermost to the outermost.

"All this constitutes the materia medica of the drug proved. Multiply it by many hundreds of drugs and we have the homoeopathic materia medica which is in the language of the people and not in medical language. It is the language of human sickness, not of pathology or tissue change.

"Provings have been repeated in order to show that repetition brings out nothing new. Once well made, the proving of a drug stands forever."

9. The Reason for using Different Potencies in Homoeopathic Prescribing

Homoeopathy is unique in that not only is a large group of remedies drawn upon, but also each remedy can be administered in any one of many different potencies. "Potency" refers to the number of times a remedy has been triturated or diluted as described at the start of this section.

The effects of giving the same remedy in differing potencies are quite different. To gain an understanding of the need for different potencies, let us consider various types of temperament and personality. We have all known the very strong, husky type of person who is only affected by strong stimuli and who is oblivious to more subtle or gentle influences. Then there are people who are affected by little things such as a draft of air, the colours of drapes or walls, a slight change in temperature. Still more sensitive are the people who can be made happy or miserable by the thoughts or facial expressions of those with whom they come in contact. These three examples represent different planes of vibration. The potency that will have the right effect on a person functioning in one plane of vibration will not have the same effect on a person functioning on another plane of vibration. Illness manifests differently in the different planes of vibration, termed the "dynamis" in homoeopathic phraseology. The more a remedy is "potentized"—that is, the more times it is triturated or diluted, the higher becomes its plane of vibration.

In classical homoeopathy it is necessary to treat illness on the specific plane of vibration in which it manifests—which means matching the potency to that vibratory plane. This cannot always be accomplished on the first try—sometimes the initial selection of a remedy may be correct but it must be tried in a number of different potencies before the right vibratory level is reached for that particular patient at that particular time. Differences in potency are much more radical that merely a difference in the "size" of a dose. "Size" loses its customary connotation in homoeopathic work because we are no longer dealing with quantities of matter as such—they have become too negligible to detect or measure. Rather, we are dealing with forms of *energy* that have been released from the matter of the drug with which the trituration or dilution starts.

Using the estimate contained in *The Twelve Tissue Remedies* by Drs. Boericke and Dewey (Philadelphia, 1951, page 28)—we find that 1/100 grain of substance contains about 16 trillion molecules. On that basis, a one-grain dose of a homoeopathic trituration above $15\times$ contains less than one molecule of the original substance or "drug" with which the trituration started. Since a molecule is said to be the smallest division of a compound, if there is less than one molecule in a dose, then by material definition the dose contains no matter of the original therapeutic substance, and the only explanation that becomes tenable for the undoubted effect produced by remedies above $15\times$ is that provided by the energy concepts as indicated in pages 93–94.

10. HOMOEOPATHIC TREATMENT OF CHRONIC CASES

A person suffering with a chronic ailment usually has a long medical history encompassing a number of ailments which have been manifest at different periods in the patient's life. To the homoeopath, the entire history is a connected sequence of events stemming from the same initiating cause. In order to accomplish a real "cure", it is necessary not only to eliminate the ailment from which the patient is currently complaining, but also the entire background of disease. This means that a remedy which actually represents the "totality of symptoms" as previously described, will have the result of taking the patient back, though often in a less aggravated form, through every illness that has been suffered through the life. Frequently a series of remedies is required, given one at a time, to retrace the illness that have occurred since birth. When the symptoms of past complaints reappear, in inverse order (that is, the most recent ones first, the earlier ones later) then the homoeopathic physician is pleased, because that indicates to him that he is on the right track towards clearing the entire body of the patient in such a manner that the chain of events which results first in one illness and then in another, is being rooted out so that a genuine condition of good health can be obtained.

To undergo the classical homoeopathic treatment for chronic ailments, is therefore sometimes a rather difficult process for the patient, requiring a great deal of patience and co-operation with the physician, including the willingness to undergo adverse reactions. Those who believe fully in classical homoeopathy are willing to undergo this process in view of the ultimate benefits that occur if the process is skilfully and correctly applied by the homoeopathic physician.

One will fail in one's understanding of homoeopathy if one regards it as simply a different method of treatment. Instead it involves totally different concepts of the nature of disease and of the nature of what constitutes appropriate treatment. For further illumination of these concepts we again quote from Dr. Julia M. Green's articles in the source mentioned, this time on the subject of:

11. "The Ideal Cure"

"If we ask any one of the so-called regular physicians what constitutes a cure, he is apt to say that the results of disease must be removed. If there is a tumour, cut it out and remove all affected tissues in its neighbourhood. If an abscess is forming, bring it to maturity and drain it. If an area is infected, open it and remove all infected tissue. If digestion is disordered and intestines clogged, use a purge. If pain is agonizing, give an anodyne, a pain-killer, and give the patient rest. If parts are inflamed, give medicine to quiet the inflammation. If there is high fever, give a medicine to lower the temperature. We might go on and on, but this is enough to show the trend and the thought behind it.

"The attempt to treat disease results in an attempt to deal with tissue changes by themselves and to ignore the man who lives in his body as his house. The trend of such treatment is to remove external expressions of disorder without curing it and so to send the expression of it deeper in toward vital centres, especially nervous centres. It is apt to create serious organic and mental troubles and render the patient himself incurable. It explains much of the increase in deep nervous disorders, focal infections, tuberculosis, cancer and insanity in these days. The basic trouble is still there, progressing insidiously towards the inner man.—

In contradistinction to the treatment of results, the ideal cure—depends on law and corollaries of law, on fixed principles which are the same yesterday, today and forever. The first fixed principle is the Law of Similars. Another is the direction of cure. We have said that the innermost man must feel the curative influence first, then his organs, tissues and so on outward to his skin, hair, nails, etc.

"There are three directions of cure: *from within outward, from above downward, from before backward*, or the disappearance of symptoms in the reverse order of their coming. Chronic disorder progresses from without inward, from surface tissues towards the vital organs to the man himself within. Cure takes the opposite direction and progresses from within outward and above downward. If the most recent ailments are the first to go and those which antedated them next to go and so on, the homoeopathic doctor knows all is well. . . .

"It is the ideal cure we are talking about. In these days most cures are far from ideal. In the first place the physician is not infallible. He is human; he makes mistakes in judgment, he does not study hard enough and continuously enough to master the details of homoeopathic philosophy and the vast homoeopathic materia medica. He gets tired so he cannot think clearly. He gets into ruts in the rush of daily prescribing.

"In the second place, the times in which we live and the environment of many patients are most difficult for the harmonious working of the homoeopathic cure. This age is intensely materialistic and given over to mass thinking and acting which is superficial and selfish. The pace of living is too fast for the gentle flow of curing described above.

"In the third place, strong drugging over a period of years has suppressed the natural expression of the patient's sickness and so has buried the very signs and symptoms the homoeopathic prescriber must have in order to select the curative remedy. A so-called drug disease is added to the natural disorder, producing the worst kind of complications and also depleting the patient's *vital force*.

"In spite of all these handicaps, it is nothing short of marvellous to see how most complicated states can be brought into order and really cured, provided the physician may have the continued co-operation of the patient.

"The ideal cure is worth striving for always."

12. THE VITAL FORCE

mentioned in the above chapter, is an important part of the homoeopathic concepts. For a description of this factor, we have found nothing that equals its presentation by Dr. Julia M. Green, so we again draw on her articles from the source mentioned:

"In thinking about the inner man expressing his ailments from within outward through symptoms, we must realize that some means of communication must exist.

"We talk about energy or vital force. Homoeopathy means by this the outflowing of the man's personality or individuality, the directing power of his soul and his mind to express himself in symptoms.

"The trained power of observation of the homoeopathic physician must be governed by law and order, the Law of Similars and its corollaries. In all government one centre rules. There may be one subcentre but one centre rules. In physical man, the centre is the cerebrum and lesser centres the cerebellum and the spinal cord; other centres are found in ganglia of the sympathetic nervous system. Messages from the cerebrum are sent through the cerebellum to points of distribution of cerebral or cranial nerves. Messages of lower and more mechanical import are sent from the spinal cord through spinal nerves. The sympathetic ganglia are a still lower grade, more mechanical or automatic. Yet, all are dependent on the cerebrum or brain from which outflowing messages or power make the rest possible.

"So, let us make a first trio: cerebrum, cerebellum, spinal cord.

"Now, let us think of the same order or outflowing from centre to circumference in some other expressions of basic principles. We have said that the inner man is really his will and his understanding and that the outflowing from these is by means of vital force; hence a second trio:

"Will and understanding, vital force, physical body.

"Another way of saying this:

"Interior man, simple or formative substance, material substance.

"We all know another trio from studying cell life. Each cell has:

"Nucleus, cell substance, envelope.

"Such thoughts as these will help us to realize the formative order in all things and to understand better, health and sickness.

"In all these views of our subject, we have a centre and a circumference with an outflowing substance between, always outflowing from the centre. Disturbances of the centre are immediately felt in all parts, even slight disturbances. Disturbances on the circumference are negligible in the healthy but may become very serious in those in disorder. In a healthy man, cuts and bruises heal quickly, leaving no disturbance. In man in disorder, results may become so serious as to cause so-called "blood poisoning". The same is true of bites of insects, of ivy poisoning, of contagious diseases. It is a question of susceptibility causing reactions which vary from mild ones to those most serious.

"In contagious diseases there is the so-called prodromal period between the time of exposure and the first appearance of the symptoms. To the homoeopathic physician this means that the centre is disturbed at the start and it takes varying lengths of time for this disturbance to manifest through the vital force to the physical body to appear as symptoms we mortals can detect.—

"The vital force, the immaterial substance through which disturbance becomes manifest in physical symptoms, can be affected only by things as delicate and subtle as it is itself. Many of the symptoms of a disturbed vital force are veiled, hard to discover and to interpret. When we have them we have expressions of the inner man, of the individual.

"Now, medicines with the power to cure must be prepared so as to reach these subtle outflowings of vital force on the same planes on which they manifest." (Hence, the preparation of remedies in the various potencies as has already been described.)

13. Homoeopathic Treatment in Acute Cases

While the course of homoeopathic treatment in chronic cases may require a longer period of time than other methods of treatment, since homoeopathy undertakes a much bigger job—nothing less than the complete rehabilitation of the entire individual—yet in acute cases homoeopathy, correctly applied, often produces faster results than other therapies. The 1918 'flu epidemic was the most virulent known, and had a very high mortality rate, yet homoeopaths, who prescribed not for the disease but for the patient, had much greater success in treatment and in saving lives; the statistics show a loss of only $\frac{1}{2}$ of 1 % of cases who received expert homoeopathic care. To quote from Dr. Julia M. Green on the method of handling acute cases:

"Let us consider an epidemic, say of the 'flu. If the physician writes down carefully all the symptoms of each patient he sees in this epidemic, and then compares the lists, he will find certain symptoms common to all his patients. These are mostly the symptoms of the disease, of the 'flu, the epidemic itself. Such symptoms have very little interest for the homoeopathic physician, so far as finding what is curable is concerned. Then there are symptoms expressing the patient, showing how he reacts to the 'flu. these are the important ones and they may vary much in one locality and another. The sum of all these will be similar to the provings of a few drugs which will become the drugs most useful in that epidemic, the so-called epidemic remedies. Epidemic remedies in one locality may be different from epidemic remedies in a distant locality, and somewhat different from those of a subsequent epidemic in the same locality.

"The study and method of choosing is the same, founded on the same principles; the differences lie in the expressions of patients as manifested in their symptoms. No one patient will manifest all the symptoms which have shown the epidemic remedies. Some will stress one group, some another. There will be much variety within the epidemic picture of those sick with the 'flu.—

• "People may be classified according to the medicines they need; medicines may be classified according to the kinds of people they will help. ... The patient manifests to the physician his disorder, as expressed in his inner life and from there outward. Drugs manifest to the physician their inner lives, that which is curative in them.

"Then all that remains is to adapt what is curative in medicines to what is curable in patients and do it in the best, most orderly fashion."

14.   HOMOEOPATHIC TREATMENT IN DISEASES OF THE PERSONALITY

As will be understood from the exposition of Man as a three-fold being, the expert homoeopathic physician is as much at home in the treatment of personality diseases (mental and emotional disorders) as in the treatment of physical disorders. Again we see the meticulous attention to minute detail. There are lists of remedies that have been established for each different symptom of mental or emotional aberration, based on the mental and emotional disorders that are caused in the process of "proving" the administration of these remedies to people in good health both physically, mentally and emotionally. The cure of mental and emotional diseases is accomplished by homoeopathy in exactly the same manner as the cure of physical diseases—namely, by the use of the Law of Similars (like cures like)—which means prescribing for a mental or emotional symptom a remedy which produces exactly the same symptom when administered to a person free of mental and emotional symptoms, and through the use of the principle of the "Totality of Symptoms" which we have previously described in this section.

The degree of detail necessary in the differentiation of symptoms, can best be illustrated by a specific example. Let us take the type of mental disorder which is characterized by delusions. There are different lists of remedies for each type of delusion. We shall choose for this example, the particular delusion, that a patient believes a part of his body has become enlarged. The "delusion of enlargement" requires different remedies than other types of delusion—each type having its own lists. But that is not all; to apply the homoeopathic method correctly, the practitioner must note the part of the body which the patient believes has become enlarged—whether arm, leg, head, eyes, etc.—because for each part there is a separate list of remedies indicated! (See *Diseases of the Personality* by Prof. Th. Ribot, translated and annotated by P. W. Shedd, M.D.).

Just as the mental and emotional aspects of the patient are taken into full consideration when prescribing for physical ailments, the homoeopathic physician finds that he must not lose sight of the physical aspects of mental diseases, as will be shown by the following quotation from the "bible" of homoeopathy— the *Organon* of Samuel Hahnemann, the founder of homoeopathy, who lived 1755–1843.

"Almost all the so-called mental and emotional diseases are nothing more than corporeal (bodily) diseases in which the symptom of derangement of the mind and disposition peculiar to each of them is increased, whilst the corporeal symptoms decline (more or less rapidly), till at length it attains the most striking one-sidedness, almost as though it were a local disease in the invisible subtle organ of the mind or disposition.

"The cases are not rare in which a so-called corporeal disease that threatens to be fatal—a suppuration of the lungs or the deterioration of some other important viscus or some other disease of acute character, e.g., in childbed, etc.

becomes transformed into insanity, into a kind of melancholia or into mania by a rapid increase of the psychic symptoms that were previously present, whereupon the corporeal symptoms lose all their danger; these latter improve almost to perfect health or rather they decrease to such a degree that their obscured presence can be detected only through the observation of a physician gifted with perseverence and penetration. In this manner they become transformed into a one-sided and, as it were, a local disease, in which the symptom of mental disturbance, which was at first but slight, increases so as to become the chief symptom, and in a great measure occupies the place of the other (corporeal) symptoms, whose intensity it subdues in a palliative manner, so that, in short, the affections of the grosser, corporeal organs become, as it were, transferred and conducted to the almost spiritual mental and emotional organs which the anatomist has never yet and never will reach with his scalpel.

"In these diseases we must be very careful to make ourselves acquainted with all of the phenomena, both those belonging to the corporeal symptoms, and also, and indeed particularly, those appertaining to the accurate apprehension of the precise character of the chief symptom of the peculiar and always predominating state of the mind and disposition, in order to discover, *for the purpose of extinguishing the entire disease*, among the remedies whose pure effects are known as homoeopathic, medicinal, pathogenic forces—that is to say, a remedy which in its list of symptoms displays with the greatest possible similarity, not only the corporeal morbid symptoms present in the case of disease before us, but also especially this mental and emotional state.

"To this collection of symptoms belongs in the first place the accurate description of all the phenomena of the previous so-called corporeal disease before it degenerated into a one-sided increase of the psychic symptom, and became a disease of the mind and disposition."

Hahnemann's reference in the foregoing quotation to the "spiritual, mental and emotional organs which the anatomist has never yet and never will reach with his scalpel", indicates that he was familiar with the fact that Man in addition to his physical body, also consists of higher bodies as we have outlined in the section on Chromo-therapy. Clairvoyants who can see and make detailed observations of the higher bodies of Man, have told us that the higher bodies contain counterparts of each organ of the physical body.

Chromo-therapy, and homoeopathy, are two methods of treatment which take into consideration the *whole man*—his entire constitution, and not merely his visible, physical body.

Gallavardin, a very successful French homoeopathic physician, writing in his book *The Treatment of Alcoholism*, described the regular medical practice of treating only the physical body, as "a species of veterinary medicine applied to man". Modern medical practice is no longer as deserving of this criticism, since psychosomatic aspects are coming increasingly into recognition. However, homoeopathy is unique in its development of methods for applying the same type of energy essences (produced by the process of potentizing the remedies) to all the different manifestations of human disorder, whether predominantly located in Man's physical body or in his higher bodies.

Gallavardin, in the book mentioned, lists in detail fourteen different types of people who become addicted to alcoholism, and gives the homoeopathic remedy suited to each type. He also gives lists of remedies for different types of undesirable behaviour induced by drunkenness, as for example jealousy, fury for striking, stupidity, etc. He also gives numerous case histories showing in interesting detail the exact application of the homoeopathic method to these cases.

15. COMMENT

Classical or Hahnemannian Homoeopathy is a pursuit which is ideal in theory but which often falls far short of the ideal in practice. The shortcomings stem from the following circumstances:

1. Although homoeopathy has been practised for 150 years, there are very few physicians practicing Hahnemannian Homoeopathy at the present time.

2. Of the limited number of homoeopathic physicians who are available, fewer yet have both the temperament and ability to apply the principles correctly.

3. Even of those very few who apply the principles to the very best of their ability, Hahnemannian Homoeopathy is a most difficult method to use. The skill and knowledge required are almost beyond human capabilities. Dr. Stuart Close, a prominent homoeopathic physician and teacher, writing his book *Lectures on Homoeopathic Philosophy* states that his bad prescriptions outnumbered his good prescriptions.

Nevertheless, the results obtained from the correct use of homoeopathic remedies is so excellent, that various avenues of approach have been explored in the attempt to overcome the difficulties. These approaches can be grouped into two categories:

I. The use of other concepts for the selection of homoeopathic remedies instead of the concepts of the Law of Similars and the Totality of Symptoms, so difficult to apply in practise if used on a basis of judgement alone. Some of the other homoeopathic concepts are outlined in the next three chapters of this section.

II. The use of instrumental or mechanical means to aid in the selection of the proper homoeopathic remedy—the one that will attain the best results for an individual at a particular time. This can be done in applying either Hahnemannian Homoeopathy or the newer homoeopathic concepts, and yields brilliant results when used by an experienced and capable operator. In fact, the aid provided by instrumental selection is so great as to virtually eliminate the difficulties which have plagued homoeopathy as a method of practice. No longer is it necessary to have a prodigious memory for a vast quantity of symptoms distributed between various remedies, nor a virtually super-human judgement in trying to match the right remedy to a conglomeration of symptoms exhibited by an individual case. Instead, what might best be described as a "resonance test" quickly qualifies or disqualifies any remedy tested. When more than one remedy qualifies, a comparative evaluation of the effectiveness of each remedy candidate is given. This indicates which of the qualifying candidates is the best choice.

There are two principal methods of making a resonance test, for remedy selection—the radiesthetic method and the radionic method. Both of them function on the principle of determining whether the radiation of a remedy is in harmony with, or is antagonistic to, the radiations of the person for whom the remedy is intended. When there is reinforcement between the radiations, the remedy is beneficial; when there is antagonism or interference between the radiations, the remedy is not of help, and could be disadvantageous.

L. E. Eeman's experiments with drug tests, mentioned on page 171, indicated that every remedy emits a radiation, that can be transmitted on wires and will impinge on the human organism with marked effect.

Experience in both radiesthetics and radionics has shown with compelling certainty the fact that every living organism emits its own complex of radiations.

It is the manifestation of interaction between these two types of radiations—from the remedy and from the patient, that makes it possible to achieve an advance determination of the effect of the remedy, either in reducing a disease condition or increasing the vital health of the body.

Radiesthesia refers to the use of a pendulum, or equipment in which a pendulum is incorporated or attached. Resonance or the lack of resonance is shown by types of pendulum motion, or changes in that motion.

Radionics involves the use of more extensive equipment, including a tuner panel. Resonance or lack of resonance is shown by alterations of the surface response of a rubbing plate to finger stroking. These variations produce a pronounced tactile difference in the "feel" of the surface, and with some equipment there is likewise an audible difference.

In the case of newer concepts of homoeopathy, the radionic method for remedy selection is the more versatile and advanced, since it permits tests to be made that give advance determinations of the effect the remedy will have upon a specific type of tissue or upon an organ in the body, that is specially in need of improvement. The vast range of tunings that can be used on radionic equipment, makes radionics the method of choice when therapeutic efforts are to be directed to a particular part of the body, or a specific type of tissue within the body. Determinations are also more quickly performed with radionics than with radiesthesia.

Regardless of which of the two methods is used, adoption of instrumentation for selection of homoeopathic remedies represents a tremendous forward step, putting the use of homoeopathy on a practical basis.

II.   THE TWELVE TISSUE SALTS—(DR. SCHUSSLER'S METHOD)

INDEX

INTRODUCTION

Eighty years ago, Dr. William Schussler of Oldenburg, Germany, promulgated what he termed an "abridged homoeopathic therapy", which has attained wider use than classical homoeopathy, and is considered by its adherents to have very substantial merit.

The Schussler system comprises twelve remedies—each a different mineral salt prepared by homoeopathic trituration, but prescribed in accordance with theories quite different from classical Hahnemannian homoeopathy.

Briefly, the principles of the Schussler system are:

1. Disease does not occur if cell metabolism is normal.
2. Cell metabolism is normal if cellular nutrition is adequate.
3. Nutritional substances fall into two categories—organic and inorganic. Sugar, fat and albuminous substances are organic—water and the twelve mineral salts are the inorganic substances.
4. The ability of the body cells to absorb and utilize organic nutritional substances (i.e., food) is impaired if there is a deficiency in the quantity of any of the inorganic mineral constituents of cell tissue, or if there is disordered molecular action of these mineral salt constituents. Illness is the result.
5. Adequate cellular nutrition can be restored, cellular metabolism normalized and illness eliminated, by supplying the required mineral salt or salts in homoeopathic form.

1. THE TWELVE TISSUE SALTS

are termed the inorganic components of the body; if the tissues are subjected to combustion, these salts remain in the form of ashes.

These salts are:

Calcium Phosphate	Calcium Sulphate
Iron Phosphate	Potassium Sulphate
Potassium Phosphate	Sodium Sulphate
Magnesium Phosphate	Potassium Chloride
Sodium Phosphate	Sodium Chloride
Calcium Fluoride	Silicea

It will be seen that sodium, potassium and calcium salts comprise nine of the total group of these twelve remedies. The information in the remainder of this section is abstracted from *The Twelve Tissue Remedies* by Drs. Boericke and Dewey, M.D., obtainable from Thorsons Ltd., 91, St. Martin's Lane, London, W.C.2.

2. CELLULAR METABOLISM

Experiments referred to in the book cited, have demonstrated:

"that the various tissue cells disintegrate rapidly in the absence of the proper proportion of sodium, potassium and calcium salts in the circulating fluid (blood). The normal ratio is 100 molecules of sodium, 2.2 molecules of potassium and 1.5 molecules of calcium. Any marked departure from this proportion is followed by a more or less rapid degeneration of protoplasm.

"The maintenance of a stable metabolism within the cell is due to the presence of these salts in the proper ratio in the fluid which surrounds the cell.—They maintain a type of physiological balance in the liquids surrounding the living cells and any upsetting of this proportional interrelation of the various salts may lead to physiologic disturbance and disease."

3. CELL FORMATION AND NUTRITION

As stated in the above-mentioned book, the oxygen of the air, upon reaching the tissues through the blood by means of respiration, acts upon the organic substances which are to enter in the formation of new cells. The products of this transformation are the organic materials which form the physical basis of muscle, nerve, connective tissue and mucus substance. None of these substances are present as such in the blood, but are formed *within* the tissues from the albumen. With them, the inorganic salts form combinations by virtue of chemical affinities, and thus new cells are formed. With the formation of new cells there occurs at the same time a destruction of the old ones, resulting from the action of oxygen on the organic substances forming the basis of these cells. Oxidation has, as a consequence, a breaking down of the older cells.

The blood contains the material for every tissue and cell of the body; it supplies every possible physiological want in the human economy. It does this by a process in which a portion of the blood plasma seeps into surrounding tissues through the capillary walls, thus making good the losses sustained by the cells from the various functions that occur in the living body.

Two kinds of substances are needed in this process of tissue-building, and both are found in the blood—namely, the organic and inorganic constituents, as already described. The organic substances of sugar, fat and albuminous substances serve as the physical basis of the tissues, while the inorganic substances, water and the mineral salts determine the particular kind of cell to be built up.

For example, the principal inorganic materials of nerve cells are magnesium phosphate, potassium phosphate, sodium and iron. Muscle cells contain the same, with the addition of potassium chloride. Connective tissue cells have silica for their specific substance, while that of elastic tissue cells is calcium fluoride. Bone cells have calcium fluoride, magnesium phosphate, and a large proportion of calcium phosphate. The latter is found in small quantities in the cells of muscle, nerve, brain and connective tissue cells. Cartilage and mucus cells have sodium chloride for their specific inorganic substance, and this is also found in all solid and fluid parts of the body. Hair and the crystalline lenses of the eyes, contain iron among other inorganic substances.

Whenever, in the living organism, new cells are to be generated and formed, there must be present in sufficient quantity and proper relation, both the organic and inorganic substances. By their presence in the blood, all organs, viscera and tissues in the body are formed, fixed and made permanent in their functions, and a disturbance here causes disturbed function.

The connective tissues have been found by the researches of Virchow and Von Recklinghausen to be far more important in function than was formerly realized. It is now known that these tissues constitute the matrix in which the minute capillaries carry the plasma from the blood to the various bodily tissues and return the same to the blood vessels; also the connective tissue serves as one of the most important breeding places of young cells, which are capable of developing out of the embryonic latent forms to the most differentiated structures of the body.

4. The Nature of Disease

Health can be taken as the state where there is normal cell functioning; that is, when by means of digestion of food and drink, recompense is made to the blood for the losses it sustains by furnishing nutritive material to the tissues; this compensation being made in the required quantities and in the appropriate places, and with no disturbance of the normal molecular motion. Only under these conditions can new cells be properly built and the destruction of old cells proceed normally, and useless material eliminated from the system.

Every normal cell has the faculty of absorbing or rejecting certain substances. This property is diminished or suspended when the cell has suffered a loss in one of its salts in consequence of any irritation. If the altered cells regain their integrity by recovering their loss, they can again perform their normal functions, and bring about the removal of morbid products, exudations, etc.

Virchow states that disease is an altered state of the cell—hence the normal state of the cell constitutes health. The constitution of the cell is determined by the composition of its nutritive environment exactly as a plant thrives according to the quality of the soil around its roots.

The altered state of a cell which constitutes disease, is not necessarily due to a deficiency in the intake of any of the inorganic minerals; instead it may be due to a disturbance of the molecular action in the body, of one or more of the twelve mineral salts. For example, in some children suffering from rickets there is a lack of calcium phosphate in the bones; the quantity of calcium phosphate intended for the bones fails to reach its goal, and has to be excreted by the urine in order to avoid excessive accumulation of this mineral in the blood.

5. The Treatment of Disease by the Use of Cell Salts

The diseased cells have suffered a physical alteration, which precludes the entrance of the required tissue salt in the form normally supplied by food. However, it is believed that by supplying the needed mineral salt in the attenuated, homoeopathic form, an energy typical of the mineral salt is introduced into the system in such a manner as to repair the cells which have suffered for lack of the required tissue salt, thus restoring the cell's ability to absorb this element from the food intake.

In agricultural chemistry, we add as fertilizer the element most lacking in the soil. In this abridged homoeopathic method (termed the "bio-chemic method") the same principle of supplying the lack is followed. It is stated that cellular equilibrium, nutrition, metabolism and function are restored by supplying the deficiency. This bio-chemic method is aimed at aiding nature in her efforts to cure, by supplying the natural remedies lacking in certain parts of the body—the inorganic salts, and in this way correcting the abnormal states of physiological chemistry.

The aim of biochemistry is to remedy a deficiency directly, with homogenous (identical) material, in contrast to Hahnemannian homeopathy which often makes use of remedies heterogenous (harmonious) to the constituents of the human organism.

6. Reasons for the Administration of Cell Salts in Homoeopathic Form

(*a*) *Minimal* (*Homoeopathic*) *Doses Correspond to the Proportion of Cell Salts normally found in the Body Tissues of Healthy Organisms.*

Boericke and Dewey in the book mentioned, quote a table from Bunge's *Textbook of Physiological and Pathological Chemistry* showing the amount of mineral salts in one litre (about one quart) of blood cells, and also in one litre of intercellular fluid (plasma), as follows:

Mineral Salt	*Blood Cells*	*Plasma*
Potassium sulphate	0·132	0·281
Potassium chloride	3·079	0·359
Potassium phosphate	2·343	—
Sodium sulphate	0·344	1·532
Sodium chloride	—	5·545
Sodium phosphate	0·633	0·271
Calcium phosphate	0·094	0·298
Magnesium phosphate	0·060	0·218
Iron phosphate	0·998	—

Figures in the above table represent grams (or decimal-fractions of a gram) mineral content per litre. These quantities range from about 1 part in 20,000 up to 1 part in 200. This represents, in homoeopathic trituration range, the 2x to 5x potencies. It is in low potencies of 3x and 6x that these mineral salts are usually administered, and sometimes 12x.

(*b*) *The Quantity Supplied must be in Accordance with Need if Best Results are to be Obtained*

When a few grains of magnesia are sufficient to cover the required daily supply how minute must be the dose of magnesia to be given for a neuralgia which is caused by an inconceivably small deficiency of this salt in a minute portion of the nerve tissue.

The amount of iron per quart of milk is about 4 ml., and a child nourished upon milk receives only about 1 ml. or 1/65 g. of iron at a time. If 4 ml. represent the daily supply of iron for the nourishment and growth of the child, distributed to all the iron-bearing cells of the body, how small must be the dose therapeutically considered, of a salt of iron to be given to allay a molecular disturbance occurring in a small part of the body, as for example an area in which there is inflammation!

(*c*) *The Use of Very Small, Diluted Doses results in Absorption by the Body rather than Excretion*

For example, if it is desired to introduce Glauber's salt (sodium sulphate) into the blood, it cannot be done by giving this salt in a concentrated form, for in that event the intestinal tract is irritation, resulting in diarrhoea which eliminates the salt from the system. But, a homoeopathic dilution of this salt will enter the blood and intercellular fluids from the tissues of the mouth and esophagus.

Large doses of iron given in the attempt to cure anaemia, have the effect of irritating the stomach and often pass out in the excretions without having been absorbed by the body. Absorption is assured when homoeopathic dosage is used.

Every bio-chemic remedy must be sufficiently diluted, according to this theory, to avoid destroying the function of healthy cells, and to restore disturbed function wherever present.

(*d*) *Minimal Doses Facilitate Absorption into the Body Tissues instead of Having the Therapeutic Substance destroyed by Stomach Secretions*

A mineral when it reaches the human stomach, is acted upon by the muriatic acid contained in the gastric juice. If the mineral be a salt of iron, then the chloride will be formed through the action of the acid. Therefore if it is desired to administer the phosphate of iron for therapeutic purposes, it must be kept out of the stomach. For this purpose a minimal (homoeopathic) dose is required, which dilutes the molecules in such a way that the substance will penetrate the tissues of the mouth, throat and esophagus and reach the blood through capillary walls. It has been found that those mineral salts that are insoluble in water must be triturated to at least the 6x potency to enable them to be absorbed by the body tissues; those soluble in water may penetrate the epithelial cells in lower potencies than 6x.

Common table salt can be readily tasted, but the particles are too large to be absorbed directly into the body cells. Therefore people who have plenty of salt in their diet may at the same time be deficient in sodium chloride, and this deficiency is readily remedied by administering salt in the homoeopathic trituration.

7. Evidences of the Effectiveness of Administration of Minute Quantities

(*a*) Light, itself an imponderable, is yet able to induce chemical reactions in plants, by which carbonic acid is decomposed into carbon and oxygen. Light induces an electro-chemical effect upon the retina of the eye and indirectly upon the optic nerve, which makes vision possible. Similarly, light exerts a chemical effect upon photographic plates and film, making possible the entire science of photography.

(*b*) Atropine, even when diluted more than a million-fold, produces dilation (enlargement) of the pupil in Man and even in the lower warm-blooded animals.

(*c*) The addition of corrosive sublimate in as small a quantity as 1 part in 800,000 (corresponding to about a 6x homoeopathic trituration) to a grape sugar and yeast solution, induces powerful fermentation far above normal, according to the findings of Prof. Hugo Schulz in the "Berliner Klinische Wochenschrift", November 4, 1889.

(*d*) The inorganic substances which serve plants for nutrition are taken up by them only in minimal quantities. Liebig, in his chemical letters, observes that the strongest manure of earthy phosphates in a coarse powder has far less effect than that of a much smaller quantity that has been finely divided, which by its subdivision can be diffused into the soil.

(*e*) Hydrochloric acid, when diluted a thousand-fold with water, easily dissolves fibrin and gluten at body temperature, but this solvent power is lessened if the proportion of acid in the solution is increased.

(*f*) The mineral waters found to have the greatest curative effects contain minerals only in extremely diluted proportions—in fact, these proportions correspond to the range of 5x to 8x potency of homoeopathic trituration.

(*g*) A striking illustration of the physical effect of very tiny particles, is furnished by the Kirchoff and Bunsen experiment in which 3 mg. of common salt (less than 1/20 of a grain) are blown into a room containing 60 cubic meters of air. In a few minutes, sodium lines appear in a flame standing at a considerable distance from where the salt was released. This represents a dilution of about 1 part in 20 million.

(*h*) Darwin, in his work on Insectivorous Plants, commented on the fact that so inconceivably minute a quantity as one twenty-millionth of a grain of ammonia phosphate induces changes in a gland, sufficient to cause a motor impulse to be sent down the whole length of a tentacle, inducing movements through an angle of about 180°. The quantity of one twenty-millionth of a grain is far less than the quantity of mineral present in a 1-grain dose of a 6x homoeopathic trituration (1 to 1 million dilution).

8.　Condensed Summary of Functions of the Different Cell Salts

Following are brief summaries of the functions performed by the different mineral cell-salts and the effects of deficiencies of these salts, condensed from the medical literature on the subject.

CALCIUM FLUORIDE is found in the surface of the bones and in the enamel of the teeth, and is an important constituent of elastic fibres and of the skin. Elastic fibres are found in the skin, in the connective tissues, and in the vascular walls.

A disturbance in the equilibrium of the molecules of calcium fluoride in the body, causes continuous dilation or chronic relaxation of elastic fibres and to render these tissues incapable of absorbing discharges, with the further consequence that the products of discharges harden in the tissues and can cause pathological enlargements of blood vessels—leading to haemorrhoids, varicose and enlarged veins, vascular tumours, hardened glands, uterine displacements and weakening of abdominal walls. Deficiency of calcium fluoride is also an important factor in malnutrition of bones and teeth.

CALCIUM PHOSPHATE is found in the blood corpuscles and blood plasma, and in connective tissue, bones and teeth. This salt gives solidity to the bones, and combines with albumin for the benefit of all tissue processes that require albuminous substances. It plays an important role in the formation of new blood corpuscles and promotes the growth of all cells.

It has restorative power after acute diseases, is of value in combating senility, and is helpful in skin diseases marked by crusts or scabs. It combines the alterative and nourishing properties of calcium with the nutritive and stimulating properties of phosphorus.

CALCIUM SULPHATE is present in the liver, particularly in the bile. It is stated to play an important role in the destruction of worn and useless blood corpuscles, through absorption of moisture from those spent blood cells. If, through deficiency of calcium sulphate, the old blood cells are not destroyed or taken up by the lymph system, then these useless blood cells reach the mucus membranes and skin, producing catarrh, eruptions, pus, abscesses, etc. Calcium sulphate is particularly helpful in treatment of infections which continue to discharge or ooze pus or other matter.

IRON PHOSPHATE is found in the haemoglobin or colouring matter of the red blood corpuscles. Iron and its salts have the property of attracting oxygen. The iron of the blood corpuscles takes up the oxygen from the inhaled air. This is carried to every cell throughout the entire body, by means of the mutual reaction of iron and potassium sulphate.

A disturbance of the equilibrium of the iron molecules in the muscular fibres causes relaxation; the resulting effect depends upon the type of muscle fibre that is involved.

Occurring in:

Muscular coats of blood vessels	Causes dilation of vessels and accumulation of blood in vessels, causing congestion. If the congestion becomes severe, haemorrhage results, through rupture of the walls of the vessels.
Relaxation of muscular walls of the intestinal villi	Results in diarrhoea.
Relaxation of muscular walls of the intestines	Results in reduced peristalsis, and constipation.

When muscular fibres need to be strengthened, as for example after an injury, iron phosphate plays an important part. Through its power of attracting oxygen, iron becomes a useful remedy in diseases of the blood such as anaemia, chlorosis, and leukaemia. Through its capability for relieving congestion, it has also been found very useful in cases of fevers and inflammations before the stage when pus and other waste matter is exuded.

POTASSIUM CHLORIDE (should not be confused with potassium chlorate, a poison)—is found in blood corpuscles, muscles, nerve and brain cells, and in intercellular fluids. If the cells of the skin (for example, in consequence of irritation) lose molecules of potassium chloride, then fibrin is thrown off which in drying, becomes a mealy eruption. If the irritation extends to the tissues below the skin, then both fibrin and serum will exude, and the involved part of the skin will blister. If potassium chloride molecules are supplied, then the affected tissue is restored to normal. It has been found very useful in the second stage of inflammatory conditions where there are exudations.

POTASSIUM PHOSPHATE, while found in all animal fluids and tissues, is especially prominent in the brain, nerves, muscles and blood cells. It is indispensable to the formation of tissues. It is the main inorganic salt involved in:

The oxidation process
Change of gases in respiration
Various chemical transformations in the blood
Saponifying of fats.

Potassium phosphate is the predominant mineral salt in the serum of the muscles, while sodium chloride is the predominant mineral salt in the circulating blood.

Its deficiency results in weakness, loss of mental vigour, depression, degeneration, neurasthenia, atrophic conditions in older people, excessive body odour.

POTASSIUM SULPHATE is the carrier of oxygen. The oxygen taken up by the iron contained in the blood corpuscles is carried to every cell of the body by the reciprocal action of potassium sulphate and iron. Every cell requires for its growth and development the vitalizing influence of oxygen.

Skin and epithelial cells poorly fed with oxygen, loosen and scale off freely. If oxygen is brought to the affected parts by means of potassium sulphate, the formation of new cells is brought about, and the damaged cells are cast off.

Another effect of deficiency of potassium sulphate, is the discharge of yellow or greenish matter and watery secretions from any of the mucus surfaces. Such secretions often occur in the third stage of inflammation. It will be noted that iron phosphate is often the cell-salt remedy for the first stage of inflammation; potassium chloride for the second stage of inflammation, and now potassium sulphate for the third stage of inflammation.

MAGNESIUM PHOSPHATE is a constituent of muscles, nerves, bone, brain (especially the grey matter), spine, sperm, teeth, and blood corpuscles. Dr.

William Schussler, the originator of the cell-salt remedy system, states that the action of magnesium phosphate is the reverse of that of iron. By functional disturbance of the molecules of iron, the muscular fibres relax unduly; through the functional disturbance of the magnesium molecules they contract excessively, resulting in cramps, convulsions and other nervous phenomena, and magnesium phosphate is the cell-salt remedy that has the most pronounced antispasmodic action.

It is also of benefit to lean, thin, emaciated persons, and to those who are weak, tired, or unable to sit up, whether from acute or chronic ailments.

SODIUM CHLORIDE is a very important constituent of the circulating fluid which nourishes the cells. Any important disturbance in the normal proportion of sodium in the body fluids, results in degeneration of the protoplasm or basic cell material of the body. It is the chemical substance found in greatest quantity in the blood plasma.

One of the main functions of sodium chloride (common salt) is to regulate the osmotic tension in the body—i.e., the extent to which fluids can seep through from one type of tissue to another. This is done by maintaining the blood serum at a particular level of specific gravity.

The function of salt in regulating the degree of moisture in cells, is accomplished through its property of attracting the water that is taken into the system in food or drink. . . . Every cell contains soda, combined with chlorine that is formed by splitting up the sodium chloride contained in the intercellular fluids. This chloride within the cell has the property of attracting water, by means of which the cell enlarges and is divided. The division of cells is necessary for the purpose of cell multiplication.

If no chloride (of sodium) is formed within the cells, then the water destined to supply their moisture is retained in the intercellular fluids, and a waterlogged or dropsical condition results. In this condition the patient craves salt, but is unable to utilize a plentiful supply of salt in the food, as the cells under these conditions cannot take up the particles of salt unless offered in a very dilute (homoeopathic) solution.

The salt in the epithelial cells of the intestinal mucus membranes, transfers the water taken with food, into the blood contained in the branches of the portal vein. A disturbance of this function through any irritation results in reverse flow, in which event serum from the blood enters the intestinal tract and watery diarrhoea results. If the irritation reaches the mucus cells of the intestines, the diarrhoea is of both a watery and mucus nature.

Another important function of sodium chloride is to regulate the quantity of water entering into the composition of the blood corpuscles, to preserve their form and consistency.

To go into still another function of sodium chloride, its molecules in the peptic glands (of stomach and duodenum) become split up by the mild action of the carbonic acid of the blood, its chlorine separated, and the free soda then unites with the carbonic acid; this combination reaches the blood while the chlorine, united with hydrogen and dissolved in water, reaches the stomach as hydrochloric acid—an important digestive secretion.

If there is a deficiency of salt in the peptic glands, then no hydrochloric acid can be formed, and the stomach secretes alkaline mucus which interferes with digestion of food. The administration of dilute hydrochloric acid for such cases is only a palliative—the cure is stated to be the administration of sodium chloride in homoeopathic form so that the molecules of this substance can be absorbed by the peptic glands, to restore the proper functioning of these glands.

If the action of salt molecules in the body is disturbed, there can exist at the same time both increased and decreased secretions in different parts of the body. For example, a gastric catarrh with vomiting of water or mucus may exist, together with constipation from a lessened secretion of mucus in the colon.

Excessive salt intake can cause profound changes, including dropsical conditions, certain types of anaemia, retention of waste matter resulting in toxic conditions, dry skin, sallow appearance, constipation or diarrhoea, obesity, abnormal appetite, incessant thirst, thinning of the blood, slow circulation and lowered temperature.

The whole quantity of salt in the human body is approximately 11 oz., and if the daily salt intake is more than sufficient to counterbalance the daily consumption in the system, the kidneys attempt to excrete the excess. This is why kidneys, if weakened by disease, are greatly irritated by excessive salt consumption.

SODIUM PHOSPHATE is found in the blood, muscles, nerves, brain cells, and intercellular fluids.

Through the catalytic action of this salt, lactic acid is decomposed into carbonic acid and water. Sodium phosphate absorbs the carbonic acid, and carries it to the lungs, where it gives up the carbonic acid and attracts oxygen to the iron contained in the blood corpuscles.

In the case of deficiency in sodium phosphate, there is insufficient conversion of lactic acid into carbonic acid, with the result that the system becomes what is popularly known as "acid", and various troubles incident to that state are likely to ensue, such as rheumatism, indigestion, intestinal ailments, etc.

Uric acid is kept soluble in the blood by the presence of sodium phosphate and the natural temperature of the blood. When there is a deficiency of this salt, then uric acid combines with soda to form sodium urate, an insoluble salt which produces crystalline deposits which can cause acute inflammatory rheumatism and even gout.

Another function of sodium phosphate is to emulsify fatty acids, and also to play a part in the utilization of albumen (protein). This mineral cell-salt is therefore of prime importance as a remedy for ailments traceable to improper metabolism of fats.

The white blood cells, leucocytes, lymph corpuscles, carry molecules of fat and peptones (modified albuminoids) from the intestinal walls to the blood and thence to the tissues. This is done directly through the walls of the intestinal tract by the leucocytes carrying peptones, and indirectly through the thoracic duct by the leucocytes carrying fat molecules, and thence to the tissues through the walls of the capillaries. Here, after the peptones are retransformed into albuminoids, they are deposited and become material for the growth of young cells which are formed by division.

If the progress of the leucocytes carrying fat molecules is stopped anywhere during the course through the lymph glands, skin, bones or lungs, then phlegm, glandular inflammations and swellings occur, and tubercular conditions are rendered possible; also fatty degeneration.

If the fatty degeneration has not proceeded too far, homoeopathic doses of sodium phosphate possess the power to free the leucocytes so that they can resume their special functions, particularly the emulsifying of fats.

SODIUM SULPHATE, along with sodium chloride, regulates the excretion of water from the cells of the body. From the opposite action of these two salts, a balance is maintained. Both have the property to attract water, but for opposite purposes. Sodium chloride attracts water for use in the system, while sodium sulphate attracts water for elimination from the system.

Sodium chloride furthers the division of cells and therefore the formation of new cells. Sodium sulphate takes water away from worn-out leucocytes so that they will disintegrate and be eliminated. Sodium sulphate has therefore been found helpful in leukaemia—an ailment marked by an excess of leucocytes.

Sodium sulphate is a constituent of the intercellular fluids, rather than of the cells themselves. It functions through irritating the epithelial cells and nerves, inducing them to send their superfluous water to the kidneys, for subsequent elimination from the body as urine.

Sodium sulphate, by stimulating the epithelial cells of the bile ducts, pancreas and intestinal canal, furthers the normal secretion of these organs. It is also able to provide necessary stimulation to the nerves of those organs.

Suppression of urine, and later the involuntary elimination of urine, can be consequences of a deficiency of sodium sulphate and hence lack of adequate stimulation of the nerves in the bladder.

Disturbed function of sodium sulphate in the body can lead to excess or deficiency of gall secretion, and hence to liver disturbances.

In cases where diabetes is due to lessened secretion of pancreatic fluids, sodium sulphate may be the remedy needed.

Some cases of constipation and colic are due to deficiency of sodium sulphate, with its power to stimulate the nerves of the intestinal tract.

Conditions of oedema (excess water in tissues) can be due to disturbed function of this mineral cell-salt, and in that event the remedy lies in administration of this salt in homoeopathic form.

SILICA is found in connective tissue, skin, hair and nails. It is a remedy for pus in connective tissue or skin, helps bring abscesses to a head, and aids in the healing of wounds. It is of value also in the treatment of bone and joint ailments. In addition, it has proved of benefit where the body is irritable and weak, and when the nervous system gives way to excessive agitation.

Silica has the faculty of absorbing bloody or serous discharges which are present by virtue of defective lymphatic action. It promotes perspiration and the elimination of internal toxins through the skin.

When the cells of any part of the connective tissue are deficient in silica molecules, they atrophy. An adequate supply of silica is therefore necessary to prevent degeneration.

The foregoing comments on each of the twelve mineral cell-salts are presented to show the great importance of these salts in bodily functioning, and are not intended as directions for treatment or prescriptions for any ailments. When using cell-salts for therapy it is very important to observe the differentiation of symptoms (as described in the section on classical homoeopathy). To present information which would be an adequate guide for the use of the cell-salts, requires a much longer essay than can be given here. The reader interested in this subject should refer to a compilation such as *The Twelve Tissue Remedies* by Drs. Boericke and Dewey. Obtainable from Thorsons Ltd., 91, St. Martin's Lane, London, W.C.2. That volume contains 450 pages on the subject, shows many more attributes of the different cell-salts, as compared to our brief summary, and goes into detail regarding their use for various symptoms and ailments. Those who wish to use these remedies in the treatment of any serious condition should seek the guidance of a physician skilled in the use of the cell-salts.

As an example of differentiation of symptoms, the book cited above lists ten different types of asthma, and for each type a different cell-salt is indicated as the remedy. There is similar differentiation for all the regions of the body and for practically all of the many different kinds of symptoms and ailments.

III. VITALIZATION OF THE TWELVE MINERAL SALTS
(DR. CHARLES W. LITTLEFIELD'S DEVELOPMENT)

Dr. Littlefield used the same twelve mineral salts discussed in the preceding chapter, but found methods for increasing their effectiveness in treatment. His researches also disclosed interesting new aspects of the significance of these minerals in the life process. Dr. Littlefield performed many experiments in which solutions were made of the different mineral salts, singly and in various combinations. His most important book, *The Beginning and Way of Life*, 1919, contains numerous micro-photographs of the actions and formations which take place in these solutions. This chapter is a condensed summary of a few of the conclusions developed in that book.

When solutions are made of the individual mineral salts in their ordinary form, crystals are formed having geometric designs—a different design for each of the twelve mineral salts.

The next step in this research was Dr. Littlefield's discovery that if the minerals were moistened with water and then the water was permitted to evaporate, a "vital force" appeared in the mineral particles. Under the microscope, when viewed in a drop of water added to the evaporated mass, mineral particles displayed motion, both independently and relative to each other; that is, the particles are observed to chase each other, attack, flee, etc., and give various evidences of life and volition. The exact techniques for observation of this phenomenon are given in Dr. Littlefield's book.

When these evaporated salts were placed in solution and viewed under a microscope (he customarily used 750 diameters magnification for all of these observations), the resulting crystalline forms were very different for each of the twelve mineral salts, than were obtained with solutions of salts that had not first been subjected to evaporation. Where before, the crystalline forms were geometrical and static, now they were radiate and dynamic—the pictures showing an identity of form with those of the different types of human tissue.

For example, the formations from "vitalized" silicic acid, in the microphotograph have the exact appearance of connective tissues. It is a well-known fact that silicic acid is a major constituent of all connective tissues. Similarly, potassium chloride, prepared in the same manner, produces formations having the appearance of lymphatic tissue; calcium fluoride produced formations typical of tendonous tissues; potassium sulphate when handled in the same manner, emitted forms identical with those of yellow elastic fibres; phosphate of magnesia—nerve tissue; iron phosphate—muscle tissue; and so on.

This becomes all the more remarkable, when it is noted that each of these salts is known in homoeopathic practice to be a prime remedy for diseases of the corresponding types of tissue. That is, magnesia phosphate has proven to be a most important remedy for disorders of nerve tissue—the same type of tissue it simulates in form when prepared in the manner described; similarly, iron phosphate has been found to have a very pronounced value in treating inflammations of muscle tissue, and it is muscle-tissue forms which are produced—on a microscopic scale—by this salt when prepared by Dr. Littlefield's method; and so on, through the list.

The correspondence between specific cell salts and specific types of tissues, is perhaps not so surprising when one realizes that the human body is built almost entirely out of the elements contained in the twelve mineral salts, plus carbon from food and nitrogen obtained from food and air.

An analysis of body chemistry shows that over 98 % of all body tissue consists of 14 elements. Twelve of these elements are contained in the mineral salts—6 positive elements and 6 negative elements:

Positive Elements	*Negative Elements*
Hydrogen	Oxygen
Iron	Chlorine

Positive Elements	*Negative Elements*
Sodium	Phosphorus
Potassium	Sulphur
Calcium	Fluorine
Magnesia	Silicon

The other elements, to make the total of fourteen, are carbon and nitrogen, previously mentioned.

The atoms of the twelve elements combine to form molecules. Every molecule has both a positive and a negative element in its composition. Thus, the twelve mineral salts are made up of the following combinations of the twelve elements listed in the positive and negative groups above:

Name of Mineral Salt	*Positive Elements*	*Negative Elements*
Calcium Fluoride	Calcium	Fluorine
Potassium Chloride	Potash	Chlorine
Sodium Chloride	Sodium	Chlorine
Calcium Phosphate	Calcium	Phosphorus—Oxygen
Iron Phosphate	Iron	Phosphorus—Oxygen
Potassium Phosphate	Potash—Hydrogen	Phosphorus—Oxygen
Magnesia Phosphate	Magnesia—Hydrogen	Phosphorus—Oxygen
Sodium Phosphate	Sodium—Hydrogen	Phosphorus—Oxygen
Calcium Sulphate	Calcium	Sulphur—Oxygen
Potassium Sulphate	Potash	Sulphur—Oxygen
Sodium Sulphate	Sodium	Sulphur—Oxygen
Silicic Acid	Hydrogen	Silicon—Oxygen

In the preceding chapter, Homoeopathy II, the *functional* importance of the twelve mineral salts for adequate cell metabolism was stressed. In the present chapter we refer to Dr. Littlefield's evidence for the *structural* importance of these substances, since the body is literally built out of these twelve materials, plus carbon and nitrogen.

However, supplying structural items is not sufficient to overcome disease caused by deficiency of these items; it is also necessary to supply them in the *form* in which they can best be utilized. For example, to use an analogy, if a wooden member of a house has broken down, such as a door or a window frame, it is not sufficient to bring in the branch of a tree to use in repair work; the raw wood must first be seasoned and refined into *lumber* before it will *harmonize* with the material of which the house was built. . . . Similarly, dumping the raw mineral elements into the system is not nearly as appropriate as to feed them in a form that comes closest to the natural form of these substances in the body itself. One step towards the approach of that form is the trituration process, for which appropriate reasons have been given in the preceding chapter. Dr. Littlefield adds the second step—the evaporation of the mineral salts before trituration, so that the very forms which they will then produce in solution are identical with the types of tissue which these same minerals build in the body.

In view of the vivifying effect of the process of evaporation upon these mineral salts, it occurred to the doctor to use salts that have been subjected to evaporation, when preparing his 6x triturations.

It will be recalled from the preceding chapter, that a 6x preparation is one that has been triturated to the sixth decimal potency—that is, it has been diluted in the ratio of one to ten, six successive times. The resulting preparation contains one-millionth part of the original substance.

The discovery of the creative or vital force which is imparted to mineral salts by the process of evaporation, thereby enabling them to form cells which then combine into replicas of the different types of body tissues, would have been a most outstanding improvement by itself. However, Dr. Littlefield went farther than that—he experimented at length with different combinations of the evaporated, vitalized mineral salts. He found that different combinations resulted in the formation, on a microscopic scale, of different types of life-forms from a variety of plants; flowers, ferns, trees, and so on—through the entire range of animal forms. This led him to postulate a basic law—that the type and characteristics of any form of matter, organic as well as inorganic, are dependent upon the grouping of the constituent elements.

By repeating the same combination of mineral elements, the replica of the same type of life-form was always produced. From his crystalline, mineral replicas of plants, he succeeded in producing seeds which sprouted into a new generation of plants, that grew flowers of different colours, all visible under the microscope. This, of course, is a type of biogenesis.

After years of research, Dr. Littlefield discovered the particular combination of vitalized (i.e., evaporated) cell salts that produces the human form. His book contains numerous micro-photographs of the life forms produced from these solutions, including the human figure. From then on, instead of administering the cell salts in individual doses, he made twelve combinations—the base of each combination was the particular grouping of cell salts which he discovered produced the human form; and the twelve combinations were formed from this base by adding to each combination an additional quantity of one of the twelve cell salts, so that the first combination had an additional amount of the first cell salt, the next combination had an additional amount of the second cell salt, and so on throughout the list of twelve.

In this way whatever cell salt was needed to make up a deficiency in the body of a patient, was administered in a combination containing all the other cell salts in the proportions in which they are incorporated in the normal human body. The selection of the salt to be emphasized was done in accordance with homoeopathic principles, depending upon the symptoms of the individual patient.

Dr. Littlefield's book *The Beginning and Way of Life*, contains numerous case histories and letters from patients, showing how rapidly this method of administering the cell salts cleared up even the most extreme and serious conditions. It was Dr. Littlefield's viewpoint, based upon his results in practice as a medical physician, that these salts, prepared and administered in accordance with the methods just indicated, were sufficient to cure all physical human ailments. Unfortunately he did not disclose, in his book the proportions of the combination of cell salts which builds the human form and which constituted the base of his cell-salt prescriptions. However, what one man has discovered, another should be able to rediscover.

Dr. Littlefield did publish in detail his techniques for:
1. Evaporating the cell salts in order to impart to them the "vital force".
2. Methods by which these salts produced living cells.
3. Methods for combining the salts to produce various microscopic life forms of both plant and animal life.

Another section of Dr. Littlefield's book deals extensively with his discovery of methods for registering mind-images (thoughts) in the mineral cell-salt solutions. He found that by mentally visualizing an image at the time that droplets of his mineral-salt solutions were on slides near a strong light, when the water has evaporated, it would be found that under the microscope the salt crystals had grouped themselves into a picture representative of the image he had visualized. This occurred only with the "vitalized" cell salts—that is, those which had been evaporated repeatedly according to his formula, and also according to his finding that abstract conceptions required a greater number of evaporations than concrete objects, if the picture was to appear in the cell salts; and high spiritual conceptions required a far greater number of evaporations, even up to a hundred or more. . . . Since Dr. Littlefield's time, psycho-kinesis, or the control of matter by thought, has become a statistically demonstrable reality, thus confirming his findings and beliefs. However, science does not yet know how to duplicate at will, the conditions which produce psychokinesis. Dr. Littlefield developed a technique using the mineral salts as already outlined, which worked successfully time after time.

It is evident that the process of subjecting the mineral salts to repeated evaporations, "conditions" them in such a manner that they become receptive to, and capable of being manipulated by, thought vibrations. From this it follows that the various types of mental, suggestive, or spiritual healing should be much more effective when the patient is fed the vitalized mineral salts, for then he will have in his body, material raised to a vibratory plane capable of receiving and registering the thought power emanating from either a practitioner, healing circle, or from the patient's own mind. In this way, a common ground, or meeting place is provided for physical and mental or psychosomatic healing.

In our section Homoeopathy I, we outlined some of the work done by classical homoeopathy in the treatment of personality disorders, emotional maladjustments and mental illnesses with Hahnemannian homoeopathy, drawing upon the wide variety of remedies used in that type of homoeopathy. In contrast, Dr. Littlefield published extensive tables on the use of his different vitalized cell-salt combinations to clear up various types of mental ailments and personality disorders. His results in this field, in actual practice, are supported by some very interesting research experiments and theoretical conclusions.

At this point, the reader should recall Dr. Littlefield's work in determining that different combinations of cell salts produce different types of life-forms. One of his discoveries was that combinations of the vitalized salts of the three sulphates (potassium, calcium, and sodium sulphates) when handled in accordance with his techniques, produce the brain-forms, as illustrated by micro-photographs in his book. Furthermore, he found that by varying the relative proportions of the three sulphates, brain-forms of different shapes are produced —with one or another lobe accentuated, and with different portions of lobes emphasized or slighted. The significance of this discovery becomes apparent only when one takes into consideration Dr. Littlefield's theory—since proven independently by extensive research in universities—that every function of the brain, not only physical functions but also every attribute pertaining to personality, behaviour and attitudes, has its own characteristic portion or location of the brain. That portion is the product of a particular chemical combination, somewhat different from any other chemical combination. If there is a deficiency in the body of one or more of the chemical-constituents required to build various portions of the brain, then certain portions or locations in the brain that are dependent upon that chemical will be under-developed and therefore their function will be distorted. The book expounds at length the basic premises that form depends upon the proper grouping of constituent elements, that function is dependent upon form, therefore the grouping of constituent elements likewise determines function.

The intelligent use of the twelve mineral salt compounds—especially in their vitalized form—is stressed as being essential, before conception and during gestation, to assure better health and well-rounded mental state and personality potentials of the individual to be born. Cases are referred to, in which the mother was too sickly to be completely restored to health before her child was born; yet by the use of these mineral compounds, the transmission of hereditary diseases to the child was prevented, and a sound, healthy child was born. This approach offers great hope for the improvement of the race, if it should be adopted on a sufficiently wide scale.

In his discussion of the problem of race improvement, Dr. Littlefield went into the much-disputed question as whether the mother could influence the personality and mental temperament of the unborn child by her own thoughts and concentration. He pointed out that during gestation, the child is nourished by the blood of the mother. If the mother's blood contains an adequate supply of mineral salts that have been vitalized by the method he describes, then a portion of these salts are absorbed by the unborn child, as has been repeatedly proven in the birth of healthy children from sick mothers treated with his preparations of the mineral salts. Since in other experiments Dr. Littlefield had already established that it is the vitalized mineral salts which have the faculty of absorbing and registering thought energy, showing that the vitalized mineral salts are a medium through which thought energy becomes translated into physical reality, he then presented the theory that by supplying the vitalized mineral salts to the mother during pregnancy, the unborn child would be supplied with the medium for receiving, registering and translating into reality the thought conceptions, ideas and desires of the mother relating to the mentality, personality and temperament of the child.

Another aspect of the book under discussion, deals with the role played by colour rays in facilitating the chemical changes which are essential both to normal metabolism and to the reception and utilization of various types of life vibrations.

Animals and man could not continue to exist on the planet, without the continuous cycle of interchange of elements between plants on the one hand, and animals and people on the other hand. From his detailed analysis of the cycles of chemical changes in the forms of the different basic elements, we shall mention here just a few points:

"The *oxygen* breathed in from the air by animals and people, unites in the body with carbon and is breathed out mainly as carbonic acid gas. This is absorbed by the plants by the *action of red and yellow rays of light*—the carbon constituent of the carbonic acid is utilized in the growth of the plant, and the oxygen liberates a large part of the oxygen into the air. By this exchange between the vegetable and animal kingdoms, the balance of carbonic acid and free oxygen is maintained in the atmosphere—the plant yielding the oxygen which the animal requires, while the animal in turn gives out the carbonic acid needed by the plant.

"*Nitrogen* is taken up by the plants as ammonia, nitrites and nitrates and is converted into vegetable protein by action of the *violet* and *indigo* rays. The resulting vegetable protein is used by the animal body in which the protein is converted into urea, uric acid and other compounds. These rapidly decompose outside the organism, yielding ammonia which the plant needs.

"Potassium, sodium, calcium, magnesium and iron unite with phosphorus, sulphur, chlorine and fluorine; and silicon with oxygen, to form the twelve salts. These all enter the vegetable kingdom from the earth, dissolved in the sap; determine by their proportion the species of plant; unite (under the influence of the rays of all of the colours) with its organic products and pass to the animal as food; for animals and man, the salts perform the offices already outlined."

Since carbonic acid is the primary compound which plants must absorb from the air, it is highly significant that the colour of most of the plant surface exposed to air is green, since green surfaces absorb the maximum amount of the colour ray complementary to green, namely red; and it is the red ray which plays the largest part in the absorption of carbonic acid bearing its load of carbon for the plant, and oxygen of which a large portion is released by the plant in purified form for re-use by the animal kingdom.

Since colour rays play a vital role in the chemical changes of elements necessary to life, it is not difficult to see how colour therapy can have a profound effect upon the physical welfare of individuals. However, colour rays are of proven value not only on the physical level of necessary chemical processes, but also to enable matter to become receptive to vibratory principles which likewise are an essential element of life.

When a person concentrates upon an idea in the presence of a mineral salt solution, and the mineral particles form into a picture of the visualized image, what effect could the person have had upon the mineral salts other than one of vibratory nature? Yet this vibration, which carries the thought energy so that it can be embodied in a physical manifestation, can only be received in the presence of light of the right colour. Dr. Littlefield always observed the mineral-salt patterns and forms in a single drop of water, by means of a microscope. Each drop breaks up the light striking it, into the seven primary colours, which spread into bands throughout the drop—the red being around the circumference and the violet in the centre, with all the intermediate colours present in ordered sequence between the extreme visible colours of red and violet. Life forms of physical substances always take shape in the red or orange portions of the drop. Likewise, fixed patterns of visualized images of physical objects always form in the red or orange portions of the drop. In contrast, fixed patterns resulting from mental concentration on abstract subjects always form in portions of the drop coloured by light-rays of more rapid vibration than red or orange, and patterns resulting from concentration on spiritual subjects form only in the indigo and violet, or central portions of the drop!

As a result of these findings, showing the effect of different colour rays in enabling the vitalized mineral salts to react to thoughts on different vibratory levels, Dr. Littlefield developed an improved type of colour-therapy lamp in which white light is conducted to chambers of water at various adjustable angles. The angle at which the light is sent through the water, determines the colour and shade of colour that will issue from the device. Thus the user, instead of being restricted to a relatively small number of colours on slides, can select any gradation or shade of colour desired; also the lamp can be adjusted to select either the ultra-violet or infra-red portions of white light which lie on either side of the visible light band. The inventor received a United States patent on this colour-lamp, granted August 27, 1918.

With this colour-therapy lamp, it was demonstrated that colour rays properly selected, can produce desired chemical changes in the human body, even to their use for the building of new tissues as in the case of severe burns that ordinarily would prove fatal under conventional therapeutic care. Examples illustrated with photographs appear in *The Beginning and Way of Life*, and other books.

To summarize, Dr. Littlefield's discoveries cover four categories:

1. *The Vital Force—a Subtle Magnetism*

This force is generated by the evaporation of water at that temperature that is normal to the given plant, animal or human form. Its demonstrated functions are:

(*a*) To enable mineral salts, when properly grouped, to develop into life forms.

(*b*) To sensitize mineral salts so that they become susceptible to mind control, and will form patterns or images of objects visualized, and also will form patterns symbolic of concepts that are concentrated upon—provided always that the salts are brought into synchronous vibration with the mental image.

(*c*) To enable solutions of individual mineral salts to crystallize into dynamic living forms which group themselves into representations of the different types of tissues present in the living organisms, in contrast to the static, geometric patterns, formed in solutions of mineral salts not vitalized.

(*d*) The use of vitalized mineral salts in homoeopathic triturations for therapeutic purposes, to achieve more effective healing and body rebuilding.

2. *Composition—the Law of Form and Function*

In the apportionment and grouping of the constituent elements of an organism, lies the cause of its form. This principle, which Dr. Littlefield demonstrated repeatedly in laboratory experiments, carries far-reaching implications regarding the origin of present life-forms. Some of these implications run contrary to certain features of the theory of evolution. The subject of evolution is discussed at length in *The Beginning and Way of Life*, but is outside the scope of this book.

3. *The Rainbow Colours—the Chemists of Nature*

Each colour has a part to play in the circulation of elements between the plant and animal kingdom, due to the faculty of different colours for inducing various chemical changes.

Chemical changes can be induced directly into the body by the use of appropriate colour rays.

Development of life-forms, or the fixing of mental images, in mineral salt solutions occurs only in the presence of the colour appropriate to the subject matter.

4. *The Fixing of Mind-images in the Mineral Salts*

A basic discovery providing physical proof of the reality and creative force of thought energies. A bridge between the physical and metaphysical worlds. Invisible vibrations made manifest in a physical modality.

IV. HOMOEOPATHIC BIO-CHEMISTRY

Unquestionably, the use of the twelve mineral cell-salts in homoeopathic practice has resulted in numerous cures, including those of many difficult and serious cases. However, in some cases the cell-salts referred to in Homoeopathy II have too slow an action, while in certain other cases the effect is insufficient. One method of increasing the effectiveness of the cell salts has just been described in Homoeopathy III, concerning the vitalization imparted to the cell salts by repeated evaporations. However, these evaporated cell salts have not been available in recent years, and another type of response to the pressure for increased effectiveness of homoeopathic remedies led to the development of homoeopathic biochemistry. This system includes many notable advances, bringing faster results and greater therapeutic effectiveness in many cases.

The principal fundamental difference between the Schussler system and later practice, is that the former is concerned mainly with supplying mineral deficiencies so as to aid in the functioning of cell metabolism and the maintenance of cell structure, while the latter is concerned also with the elements needed to activate and regulate *endocrine functioning*, to the end that the various bipolar potentials in the body will be restored to the degree required for good health. Cell maintenance and cell metabolism are likewise normalized through correction of glandular malfunction, since the role played by the glands is of vital importance in cell functioning.

Important considerations in this method include:

1. *The Use of Activating and Inhibiting Elements, with Reference to Organs and Glands*

Detailed studies have been made which resulted in the determination for each different gland and organ of the body, of the elements that activate and the elements that inhibit the action of the particular organ or gland. With this information, one is able to bring the actions of all the different glands and organs into harmonious balance with a degree of scientific precision quite remarkable to observe.

2. *A New Way to use the Principle of "Potentizing"*

After determining for example, the element needed to activate or stimulate the functioning of a particular organ, one can grind into the trituration of that element, a small quantity of the gland or organ which it was desired to treat. When the resulting tablet was taken internally, the effect of this "potentizing" was that the substance would seek the particular tissue corresponding to the organ-substance that had been used for potentizing. For example, if it was desired to treat the mammary gland, some mammary substance would be included in the trituration of elements selected for the treatment of the mammary glands. If it was desired to treat the kidneys, some kidney substance would be added to the trituration of elements having a therapeutic effect upon the kidneys, and so on. This means that instead of diffusion of the remedy throughout the entire body, its effect was localized in the area where it was needed, with far more specific and definite therapeutic effect.

3. *The Restoration of Normal Bi-polar Potentials within the Human Body*

Those who have made a detailed study of glandular interactions, and of inter-play of secretions from various different organs in the body, have been most impressed with the numerous bi-polar relationships which must be maintained at adequate potentials, if good health was to be preserved or regained. From this information, one can select special remedies to accomplish the purpose of restoring adequate bi-polarity. We shall give a few examples later in this chapter.

4. *New Information on the Causation of Various Diseases*

For example, clearing up cataracts by treating the kidneys, curing paralysis agitans by treating the para-thyroid glands, and curing lesbianism by treating the mammary glands. These are only a few examples of a vast range of know-ledge, concerning hidden causative factors of various different diseases.

5. *New Information on the Part played by Various Glandular Secretions in the Utilization, Absorption and Conversion of various Mineral Elements in the Body*

For example, it can be of vital importance to know, when treating ailments where there is an excess of sulphur, that the halogen principle of the thyroid facilitates the secretion which can induce excretion of excess sulphur from the body; phosphorus, according to recent studies neutralizes sulphur but does not cause its removal from the body. By bringing the thyroid up to par, which can be accomplished by feeding—in homoeopathic form—the trace metals and minerals that serve to activate the thyroid, excess sulphur is eliminated from the system. Iodine only activates a part of the thyroid secretion complex. It has been found there are other elements required to activate the entire thyroid complex.

6. *The Use of Homoeopathic Glandular Triturations*

Homoeopathic bio-chemistry makes use of triturations of glandular sub-stances. The pure homoeopathic glandular triturations which have been produced, used separately or in combination with other homoeopathic biochemical remedies, produce outstanding results in case after case.

When using gross doses of glandular substances, there is usually an initially stimulating effect. However, all too often the continued use of glandular sub-stances in gross form results later in progressively lessened output of the patient's own glands, with consequent need for increase in the dosage. This, in turn, can lead to further deterioration of internal glandular function, and so on.

In contrast, the use of homoeopathic dosage of glandular substances appears to have the effect of stimulating the output of the patient's own glands, rather than of creating a habitual, ever-increasing dependence upon an outside source of glandular material.

Complete information on homoeopathic biochemistry cannot be presented here, for it would require a book in itself. The following pages will contain some examples of this type of homoeopathy, which will illustrate some of the principles involved and show how advanced are its conceptions.

Notes on Methods for Treating Toxins and Infections

The experience of many homoeopaths has led them to concur in the Naturopathic conclusion that the majority of all cases of illness have as their fundamental trouble, toxic conditions and internal infections. Therefore, a great deal of time and thought have been devoted to the elimination of these conditions. One of the resulting remedies is a homoeopathic tablet containing calcium, charcoal and sulphur, potentized with mammary gland substance. This tablet has a very powerful faculty for eliminating pus and infectious matter from soft tissue anywhere in the body. It has been found that the mammary glands play a very important part in the protection against infection, and that internal mammary-gland secretions exist in both men and women. These conclusions have been verified by other research scientists. Several years ago, it was interesting for homoeopathic followers to read the front page news that medical doctors in Cincinnati, Ohio, had "discovered" that the nursing mothers' milk contained an immunity factor which prevented babies from contracting polio during the nursing period. It had been discovered by homoeopaths, many years before, that the mammary secretion provides immunity not only against polio but against all types of infection, and the preparation referred to in this paragraph was designed to extend this same immunity to adults, through reactivation of the patient's own mammary glands.

Silicea, or silicic acid, has long been used by the Schussler followers as a mineral salt that helps to eliminate pus from the body. Bio-chemical homoeopaths found that by using a slightly different form of silicea, namely silicon dioxide, prepared in homoeopathic form, one had a remedy which excelled at eliminating pus from the bony structures of the body. It has proven good practice to have toxic cases, or those with internal infections, pus or abscesses, to take the mammary-potentized tablets and the silicon-dioxide tablets on alternate hours—the one to eliminate pus, and so on from soft tissue, and the other to eliminate it from bone. This produces a complete clearance from infection in many cases, and avoids the hidden complication which has been found so often to occur with other types of care—namely, that the bones served as reservoirs of infection, usually without any external symptoms or evidence and as a result, patients would remain in a run-down, precarious condition from time to time, when actually these were only re-infections spreading from the reservoirs of infection maintained in the patient's own bones.

To restore toxic patients to health, it is important to produce elimination from the body of not only the actual infectious matter, but also of the toxins, which are self-poisoning factors resulting from the body's attempt to combat the infectious matter.

The very important role played by the lymphatic system in clearing the body of toxins has been recognized, and a homoeopathic lymph preparation has been produced which was very helpful in re-activating sluggish lymphatic systems. It has been your editor's experience, as the result of certain methods which he has had for ascertaining the performance of the lymphatics, that this system is greatly under-active in almost all cases of illness. In addition to the homoeopathic lymph tablets which have been available, a homoeopathic combination of lymph and blood salts, for many cases was even more effective than the straight lymph trituration.

Another point in the homoeopathic biochemical armament against toxicity consists of homoeopathic adrenal cortex, since one of the functions of the adrenal cortex is to liquefy the waste products from the body cells, and hence facilitate the elimination of waste from the body. When waste products are not sufficiently liquefied, the cells are clogged with waste which cannot readily be eliminated.

In the case of adrenal cortex also, a compound of this substance triturated with blood salts was found even more helpful in many cases than the straight adrenal cortex. In addition, the adrenal cortex helps to activate the lymph system, hence plays a double role in combating toxicity.

Another discovery was the part that is played by the parotid gland in fighting toxic conditions, so a trituration of parotid substance was made available for use when needed.

Another link in the chain of bodily mechanisms, for purifying the system, was found in the hormonal action of certain secretions of the sex organs. By separating the health-intensifying secretions from the procreative secretions, and producing separate triturations of what might be termed the "health-factors" from the sex organs, the vitalizing and de-toxifying effects from those organs can be imparted by homoeopathic tablets without stimulating the sex urge, for those cases where it is desired to avoid sex stimulation of the conventional type. This separation of factors from the sex glands has been accomplished for both men and women.

Notes on Homoeopathic Discoveries regarding Bi-polarity

A most important factor in maintaining health was found to lodge in a group of bi-polarities between different glands or organs. As examples, we shall mention a few of them here:

Men	*Women*
1. Adrenal medulla *v.* testicle	Adrenal medulla *v.* ovary
2. Brain *vs.* testicle	Brain *vs.* ovary
3. Spleen *vs.* prostate	Spleen *vs.* placenta
4. Thyroid *vs.* para-thyroids and gonads	Thyroid *vs.* para-thyroids and corpus lutem
5. Pancreas *vs.* prostate and testicle	Pancreas *vs.* placenta and ovary
6. Thyroid *vs.* adrenal medulla and thymus (same for both sexes).	

Each of these bi-polarities contains the key to one of the secrets of human illness.

The interplay between adrenal medulla and testicle in men, and between adrenal medulla and ovary in women, if faulty, results in low blood-pressure, and it has therefore been found that the cure for low blood-pressure lies in restoring those glands to full function and balance.

The interplay between pancreas on the one hand, and prostate or testicle in men, placenta or ovary in women, was found to contain the key to high blood-pressure from hardening of the arteries. The role of the pancreatic secretions in promoting elasticity of the blood vessels was thoroughly explored, together with means for correcting deficient calcium metabolism so as to stop and reverse the excessive deposit of calcium in blood vessels.

Unbalance in the triangle interplay between thyroid on the one hand and para-thyroids and sex hormones on the other, was found to be at the foundation of a large array of human ills, ranging from toxicity through disorders of calcium metabolism, to lack of vitality and disorders resulting from inadequate or un-balanced sex hormones.

Another triangular interplay occurs between the thyroid on the one hand, and adrenal medulla and thymus on the other, having for its purpose the regula-tion of the sulphur-phosphorus metabolism of the body. This is one of the most important balances in body chemistry. As one bio-chemist expressed it, "We came into this world through phosphorus, and leave through sulphur." That is, phosphorus is the key element in bodily anabolism, while sulphur is the key element in bodily catabolism. Health and prolonged life require a proper balance between the two. The thymus and adrenal medulla have among their functions that of eliminating excess phosphorus from the body, while the thyroid has for one of its functions the elimination of excess sulphur from the body. If these glands are not functioning normally, the sulphur-phosphorus balance is disturbed, with disastrous results. An excess of phosphorus causes premature disintegration of bone; an excess of sulphur causes premature degeneration of soft tissue, among many other effects.

The discovery of the interplay between brain, as one polarity, and gonads, as the other polarity, was of vital importance in providing clues to methods of curing serious brain damage, as for example from auto accidents. It has proven possible to restore to health, victims of brain injuries, which other types of care had given up as hopeless. This has been accomplished through alternate feeding of homoeopathic brain substance and homoeopathic gonadal substance, thus feeding both poles of the central axis of the human organism. This is an indepen-dent confirmation of the principle discovered by Ghadiali—that green, the colour activating the head, has the same vibratory rate but opposite polarity from magenta, the colour activating the sex organs.

Another bi-polarity, listed in our table, is a most valuable key to the process of senility. Other scientific investigators have established that the process of senility results from proliferation of connective tissue throughout the body, and have learned that the control of connective tissue-growth lies in the spleen. Homoeopathic biochemists have gone farther into the chain of causes, and discovered the following sequence:

Atrophied testicles cause enlarged prostate.
Enlarged prostate causes enlargement of spleen.
Enlargement of spleen results in excessive hardening and proliferation of connective tissues, the direct cause of senility.

A similar sequence exists in women, between ovaries, placenta, and spleen.

The Role of the Trace Metals in Maintaining or Restoring Bi-Polarities Through Glandular Activation

Included in homoeopathic biochemistry is the invaluable knowledge that the *trace metals* play a most important part in activating the different glands, that for each gland there is a particular trace metal or group of trace metals that serve as activating factors, and that by the use of these metals fed to the body in homoeopathic form, the glandular functioning can be raised to optimum levels, with consequent tremendous benefits in the elimination of illness and restoration and preservation of health. By trace metals are meant gold, silver, platinum, zinc, manganese, nickel, iron, cobalt, copper and others. Although there is an extensive literature on the use of the mineral salts, there is very little available on the use of the trace metals. A few homoeopathic pioneers have advanced the frontiers of knowledge of the functions and uses of many of these metals.

In an era where medical thinking is dominated to a large extent by the concepts of suppression—suppression of symptoms, suppression of bacteria with antibiotics, suppression of natural bodily reactions with anti-histamines—the keyword of homoeopathic bio-chemistry has been "re-activation". The body has within it, mechanisms for self-policing, for elimination of harmful bacteria, in fact for attending to every internal bodily need. Instead of trying to supplant normal bodily functions, the more natural purpose is to re-activate them. There has been vital work on the *positive* sex metals, the metals which are the keystone of health by activating the vitalizing side of the sex secretions; and also on the role of the *negative* metals in activating the thyroid. Incidentally, it has been found through homoeopathic biochemical research that much current information on the thyroid is fallacious; for example the discovery has been made by some advanced homoeopaths that conventional basal metabolism tests check not so much thyroid vitality as sex-gland vitality, and that a high basal metabolism reading can accompany a low rate of thyroid activity, and vice versa. Also that the principal function of the thyroid is to control, not oxygen metabolism as currently supposed, but rather *hydrogen* metabolism. This advanced knowledge of the thyroid, its functioning and activation, can be used to solve many a problem case which had found no help elsewhere. According to these findings, iodine is only one of a group of elements needed for complete reactivation of the thyroid—four others are also needed.

CONCLUSION

Homoeopathic bio-chemical research has extended to practically every known ailment, from the simplest—like pimples, to the most serious and complex—such as osteomyelitis and tumours. The role of homoeopathic chromium preparations, as one of the factors in the control of fibroid growths, is just one of numerous major discoveries. A compound of homoeopathic phosphates has proven effective in protecting pregnancies in women who had had persistent tendency to miscarriages. A simple mineral-salt preparation has brought excellent results in dissolving blood clots in the heart and brain. These and associated methods have resulted in perfecting numerous "specifics", and this marks a departure from classical homoeopathy. Hahnemann and his successors developed homoeopathy because of the failure of specifics of former times to effect the necessary cures, in so many cases.

Classical or Hahnemannian homoeopathy treated the patient rather than the illness, but proved very difficult to administer properly, unless instrumentation was used to aid in the selection of remedies—the main reason for the decline of the old-fashioned type of homoeopathy in recent decades. Homoeopathic biochemistry represents another swing of the pendulum, in the development of more effective "specifics" in homoeopathic form, through new concepts which have led to greater understanding of the manner in which the human body functions, and the factors which underlie the various types of illness.

THE BIO-PHYSICAL RESEARCH OF DR. GEORGE A. WILSON

(Life Energies from Foods, as measured Electronically)

THE BIO-PHYSICAL RESEARCH OF
DR. GEORGE A. WILSON

THE NEURO-MICROMETER

ABOUT four years ago, Dr. Wilson completed the initial development of a very sensitive electrical device termed the Neuro-Micrometer, which registers on an electric meter-scale the flow of nerve electricity as picked up by electrodes brought into contact with the body surface. This instrument, with its many stages of amplification, has a number of different circuits for the measurement of:

Nerve energy flow	Cell oxidation
Reserve energy	Toxicity
Functional energy	Nerve pressure
Acidity and alkalinity	Emotional tension

This instrument is essentially a more advanced, more stable and refined version of the type of instrument of which the Ellis micro-dynamometer was a forerunner. The Ellis instrument has been used in certain doctors' offices for years.

With the neuro-micrometer, an approximation of the condition of individual organs or glands, can also be obtained by placing the pick-up electrodes at reflex spots on the body surface corresponding to those organs.

Sales of the neuro-micrometer have been slow, as the Wilson group has been more interested in research than in the immediate profits that might accrue from pressing for volume of sales. In the following pages, we shall briefly summarize some of the concepts which have been developed through factual measurements made by the neuro-micrometer findings.

THE CALORIC THEORY

Dr. Wilson shows the fallacy in the idea that calories both warm *and nourish* the body. In actuality, they do warm the body, but they do not nourish it! And, calories do not produce energy in the body.

Because calories create energy in mechanical devices, an erroneous concept has arisen that they also create energy in the human body, but Dr. Wilson has shown that this is not true; that humans have no mechanism for transforming calories into energy. He proves this by measuring the flow of neuro-electricity in the body before and after meals composed of various different types of food. In fact, he finds that an accelerated intake of carbo-hydrate foods forces the body into a greater *expenditure* of energy in order to burn up the carbohydrates to produce heat—this greater expression of energy has been mis-interpreted to mean that more energy has been created, whereas in fact the reverse is true—i.e. the body has lowered its reserve supply of energy. Thus the caloric theory of diet is based on a false premise, and has occasioned much erroneous thinking on the subject of diet.

As Dr. Wilson states—For a health system to be worthy of its name and to justify the confidence placed in it, such a system must be built around the correct interpretation of what happens to foods in the body on the one hand, and on the other it must be built on a correct understanding of the nutritive values of foods and what it is that creates such values. That is the aim of this book.

CLASSIFICATION OF FOODS

Sugar and starch foods (carbohydrates) and fat foods (hydrocarbons) have the same elements, but combine differently in the molecules. Fat is stored fuel— not energy fuel, but fuel through which to heat the body at some future time when other fuels do not happen to be available, as happens during starvation or fasting. People can put on fat through ingestion of sugar and starch foods—the carbohydrates, as well as through consumption of the fats or hydrocarbons. Carbohydrates and hydrocarbons maintain the *thermal* integrity of the body.

Carbohydrates and hydrocarbons are made up of carbon, hydrogen and oxygen; *proteins* are composed of the same elements plus nitrogen and sulphur. Thus it is evident that proteins can take the place of carbohydrates, but carbo- hydrates cannot take the place of proteins. Some object to the nitrogen in proteins, yet the cells of the body are largely formed of nitrogen and would suffer if no nitrogenous foods were eaten. Proteins maintain the *structural* integrity of the body.

Mineral salts are necessary for the extraction of energy from other foods; deprivation of mineral salts while maintaining a diet adequate in all other respects (in animal experiments) results in depletion of nerve energy until death ensues, even though autopsies show that the body structures were all well nourished, and that the food had been properly "digested" according to con- ventional standards.

WHAT IS HEALTH?

"Health is best expressed in terms of functional energy, and it is through food that the creation of this energy is made possible. When this is true, food must be made to serve a four-fold purpose; maintain the thermal, chemical, structural and functional integrities of the body. All four of these are important, but in the past only the first three have been emphasized or given consideration. Therefore food in its relationship to the functional activities of the body has been neg- lected."

One of the main purposes of the Wilson research is therefore to explore this relationship. Other purposes are to learn more about the physiodynamics of life and the body's power of response—factors that are becoming of increasing importance in understanding and correcting disease.

THE ENERGY-FACTOR SYSTEM OF DIET

We have said that health is expressed in terms of body energy—i.e., functional energy, which is a form of electricity, called neuro- or bioelectricity, and does not come from carbonaceous foods.

Electricity is defined as a "wave of electronic motion set in action by electrons being forced to move from atom to atom". In electrical engineering, nine different methods for forcing electrons into motion are recognized, and enumerated in Dr. Wilson's book. Of these nine methods, only one is of significance in our exploration—that of chemical reactions (such as a battery charge), which in the body is expressed by the term "cell oxidation".

The basis for the creation of nerve energy is a chemical reaction brought by cell oxidation.

"Cell oxidation takes place as a result of oxygen and nutriment being absorbed into the cells. As the oxygen and nutritive elements reach the nucleus of the different cells and become a part of the particular type of chemistry of their nuclei, a minute electric charge is created from the combined action of these chemicals on the oxygen, bringing about an oxidative process. The amount of the charge thus created is about a 50-millionth of a volt. While that is not very much, it is sufficient to maintain the life-activity of a cell; and when combined with that from other cells, as happens in the brain, is sufficient to cause a flow of nerve energy that, because of its being directed towards a certain part to produce action, is called an impulse."

THE NEW CONSIDERATIONS IN DIET

The prime consideration in diet should be the effect it will have upon the cells of the body, in aiding or abetting their creation of nerve energy. Dr. Wilson enumerates four ways in which interference with the functions of cells can occur through incorrect diet:

1. Failure to supply the correct nutriment to cells, needed for cell oxidation and for the reproduction of new cells.
2. Supplying foods which create too much acid, thus interfering with cell oxidation.
3. Supplying foods which do not create sufficient acid, needed to maintain the body's unequal polarity and dissimilar potentials which are essential for causing the flow of nerve energy.
4. An excessive quantity of the types of foods which clog the fluid circulation of the body; this prevents the nutriment from reaching the cells in quantities adequate to maintain proper cell nutrition and also means there is

inadequate elimination of waste material from cells. Both of these effects produce cell malnutrition, the one basic cause which Dr. Wilson finds is involved the most in sickness, especially in chronic ailments.

FACTORS GOVERNING THE FLOW OF NEURO-ELECTRICITY

By the use of the neuro-micrometer, much basic information regarding the behaviour of nerve energy in the human body has been discovered. Following are some of the main considerations resulting from this research:

The body or any part of it when in a state of rest requires little nerve energy, for its functional activity is at a minimum, so only a little nerve energy flows. As the rate of activity of the body or any part of it is increased, the need (demand) for nerve energy increases, and if the person is in a state of good health, the demand for additional flow of nerve energy is met by an acceleration of the quantity or rate of flow.

Dr. Wilson views the amount of nerve flow occurring to any part of the body at any given time, as the result of the *demand* of that part of the body for a given quantity of nerve energy, and the *response* of the nervous system to that demand.

An interference to nerve flow, decreases the amount of energy that can be supplied to the part involved, thus impeding the normal response to a demand. Then sickness develops. While health, which is based upon a normal response to all demands, is supposed to be automatic, it can only be maintained to the extent that the nerves are free to carry the energy necessary to each part which will enable it to successfully meet all demands.

Therefore the law of nerve flow is as follows: The flow of nerve energy to any part of the body takes place only because of a demand and in relationship to the demand, as modified by interference to its flow.

Whatever is done to the body, as for example, excitement, activity, or even eating a meal, its first effect is to pull down the body's reserve and functional energy. These two types of energies are measured separately on the neuro-micrometer. One of the surprising discoveries from the use of this instrument is that the process of digesting a meal first has a depleting effect upon the body energies, before the digestion process has proceeded far enough to create new energy for the body. Another surprising discovery is that after certain types of meals, the functional and reserve energies do not recover to the point or level prevailing in the body before the meal was eaten! THIS MEANS THAT THE FOOD EATEN DID NOT MAKE POSSIBLE THE CREATION OF THE QUANTITY OF ENERGY REQUIRED FOR ITS DIGESTION!

Again it is helpful to refer to the analogy of the storage battery. When starting a car, the battery's charge is depleted and stays depleted until the generator takes over and builds the charge back up. Similarly with the body, which has only a minor amount of energy available at any given time, stored in the nerve plexuses. The demand (as for digestion of food) to which the body responds, pulls down its functional and reserve energies, which later normally is built up. The length of time required to build the charge back to normal, depends upon the body's power of response, and this power of response is found to be largely determined by the nutritive elements ingested. So both the *extent* and the speed of restoration of bodily energy-charges is dependent upon the types of foods eaten.

The flow of any type of electricity, including neuro-electricity or nerve energy, can occur only when there is unequal polarity between the terminals of the path along which the flow occurs. The extent of flow depends upon the difference in potential between the two terminals. Therefore "the body is a bio-electrical, bi-polar mechanism of unequal polarity and dissimilar potentials which, when either of these is changed, becomes sick; and ceases to live when they become equal or lose their dissimilarity."

The controlling bi-polarity has been found to be that of the brain as compared to the rest of the body—it is the brain cells which create the energy necessary to co-ordinate the functional activities of the body. When body poisons—toxins, excess acids or alkalines, prevent the cells of the brain from creating the necessary energy for such co-ordination, then there is no more energy in the brain than in the different other parts of the body; the bipolarity has been destroyed; both poles have become equal, and when they become equal, the body is like a battery which when its poles become equal, is dead.

Nerve flow takes place only when the potentials of the brain and other parts of the body are dissimilar—and in fact the more unequal, the better. Any change in the body's bi-polarity or in the extent of its neuro-electric potentials, will cause sickness or death in the degree to which the cells have been poisoned. It has been found through this research that the neuro-electric potentials are lowered and the bi-polarity is reduced, only through the effect of cell poisons upon the brain. Health and sickness are dependent upon the conditions of the cells of the body, and diet is an important factor in the condition of the cells.

Certain people who are known to have serious ailments, live in a comparatively good state of health for a long time, while others who have less serious ailments have their lives cut short. Measurements with the neuro-micrometer provide the answer to this seeming paradox; it is found that it is the poison from an ailment, rather than the ailment itself, that causes sickness or death. In some individuals an ailment creates poisons faster than the body can handle; while in others the poisons are created more slowly and at a rate which the body is able to handle. Clearly, any study which points the way towards a reduction of body poisons, either through diet or other means, carries great potentialities in therapy.

There is still another factor to be considered, in arriving at an understanding of nerve flow—and that is the ability of the nerves to transmit an electric charge throughout their length. For, the application of unequal potentials to the ends of a nerve do not assure the flow of nerve energy unless the nerves are capable of carrying the flow. It is found that it is the mineral elements in the nerves which take the neuro-electric charge and convey it to its destination through the medium of the nerves. Therefore the ability to carry nerve impulses satisfactorily, is dependent upon an adequate supply of mineral elements. If the intake of mineral elements or salts into the body is inadequate, there will be insufficient nerve flow to meet the larger demands for neuro-electricity, the demands invoked by activity, excitement, digestion, etc. In this, diet is of prime importance.

In a hydrogen ionization, the mineral salts become good conductors, because such an ionization causes the body fluids to become slightly acid. But when the ionization is hydroxyl, as happens when there is a low cell virility, the body will be too alkaline and a poor conductor. Therefore diet, through which the body should be supplied with mineral salts and the proper acid–alkaline balance, is of great significance in its relationship to nerve flow.

THE BODY'S POWER OF RESPONSE

It has long been known, that some patients have a much greater ability to recover from illness than do others. Various terms have been applied to this phenomenon, but the underlying reasons have been a mystery. Dr. Wilson terms the ability to recover from illness, as "The Body's Power of Response" because:

It is the body's response to a demand.

It represents the degree of the body's power to meet a demand.

With a neuro-micrometer, the doctor can determine quantitatively, the extent of the body's power of response at the start of any case. The primary factor in this determination is the measurement of reserve energy. It has been found that when this factor measures 90 to 100, a quick recovery is possible, if adequate treatment methods are used. Between 80 and 90, a fairly large number of people will recover from illness. As the readings of reserve energy drop below 80, the chances for recovery progressively diminish, until those cases which register below 60 invariably are found to be hopeless to save, unless the reading has merely been depressed below 60 by some heavy strain of a temporary nature as from overwork or from becoming severely chilled.

Another factor in determining the body's power of response, is that of functional energy, which also can be read in units on the neuro-micrometer.

Food selection has been found to be a most important factor in building up the body's power of response, and therefore, its ability to recover from illness. From the investigation into the effects of different types of meals as shown by the readings of the different factors registered on the neuro-micrometer before meals and at a series of intervals after meals, an entirely new system of diet has been originated. In this system, the factors are viewed from the standpoint of the effect of food in creating or favouring the flow of neuro-electricity, which, as we have found in this presentation, is a most important key to functional health.

These factors are:

1. Life-Energy factors.
2. Acid-Alkaline factors.
3. Mineral-Vitamin factors.
4. Bulk factors.

LIFE–ENERGY FACTORS

Dr. Wilson points out that: "Studies of foods by dietitians up to the present time have been largely made on the basis of the contributions foods have made towards maintaining the structural, chemical or thermal integrity of the body. Thus food has not been studied in relationship to its contributions to the body's functional integrity (energy-flow relationships). True, they have given us the caloric system but as we have found, it is based upon a wrong premise." Calories supply heat rather than functional energy. It is important to keep the body warm, but it is also important to supply the energy required to carry on its functional activities. Since this phase has been neglected until its recent discovery by Dr. Wilson, the question of supplying the body with functional energy has been left to chance.—"But the body can no more create the full amount of energy it needs out of wrong foods than it can build a sound structure, supply adequate chemicals for its secretion or maintain the correct body temperature, when eating is left entirely to chance."

The discovery, when using the neuro-micrometer, that meals composed of certain types of foods resulted in a satisfactory recovery of bodily energies while meals composed of other types of foods left the body depleted of energy, opened the way to a whole new field of research. The Hemoelectrometer was developed, which was capable of registering the electric charge of 50 milligrams of dehydrated food in 150 ccs. of distilled water. (Dehydrated foods were used for this checking, because of the necessity to obtain control over the variable factor of water content.) From this research, tables were compiled giving the life-energy figures for each different type of food. Many of these tables are published in Dr. Wilson's book *A New Slant To Diet*, obtainable from Standard Research Laboratories at E. 10th Ave. and Jersey St., Denver, Colorado, for $6.00.

Here we can only summarize the results briefly. Refined foods checked very low in life energy. Wheat in its natural state registered 24 units; in its refined state only 3 units. Fruits registered between 14 and 47 units. The leafy vegetables registered the highest in life energy—especially the leafy portions. For example, red beets registered 39 units, beet leaves 106. Celery roots 41, celery tops and leaves 97, etc. *It became evident that foods vary in life energy in accordance with their exposure to sunlight and their ability to absorb the effects of the sun's rays!*

Investigation showed that what was being measured in foods by this method, was their degree of ionization. This is highly significant in relation to the flow of neuro-electricity, since that flow is only made possible by an adequate supply

T.E.—L

of hydrogen ions which are positively charged particles of dissolved mineral salts, called electrolytes. It is in the leafy vegetables that the ionized mineral salts are present in the greatest concentration. It would seem that our discussion of the mineral salts in the section on Homoeopathy becomes even more significant in view of Dr. Wilson's findings. Do not the leafy vegetables comprise a natural homoeopathic form of administration of the mineral salts? It would also be interesting to ascertain whether the homoeopathic trituration process increases the ionization of a substance.

When persons were fed meals planned to include ample quantities of foods high in life-energy units, measurements of the flow of neuro-electricity with the neuro-micrometer showed a consistent gain in body energies. Detailed charts are presented in Dr. Wilson's book, which as already mentioned, can be obtained by anyone interested in this subject.

NEW LIGHT ON THE ACID-ALKALINE BALANCE

Normal body acidity is necessary to the maintenance of life. Otherwise there would be no circulation of nerve energy, for a certain degree of acidity is necessary for the maintenance of bi-polarity, the dissimilarity of potentials which is a necessary requirement if there is to be the flow of nerve energy.

When an area becomes diseased or injured, an accelerated nerve flow takes place to that area, because the first effect of a disease or an injury is to increase the acidity of the affected area. The increased acidity changes the potential of that area and increases its neuroelectric conductance, which steps up the flow of energy from the brain, and along with it, causes a stepped-up flow of blood. The increased nerve and blood flow, bring greater curative power to the affected area. However, too much or too little acid upsets the curative process.

An excess of acidity in the tissues of the body as a whole, induces too great a flow of neuro-electricity and, if continued too long, starts to interfere with the ability of the brain cells to create nerve energy. Diminuation of this ability, leads toward equalization of the polar differential between the brain and the rest of the body. When this equalization becomes complete, as happens in severe acidosis, death ensues, for life is dependent upon maintaining a difference in polarity between the brain and the other parts of the body. Excessive acidity has been found to be caused almost wholly by incorrect diet, and the remedy lies in a revised dietary programme.

If the body is too *alkaline*, the mineral salts in the nerves, instead of developing a hydrogen ionization, which will take and convey a neuro-electric charge, will develop a hydroxyl ionization and the nerves will become non-conductors of neuro-electricity. The result is a lessened ability of the nerves to transmit nerve energy; this interferes with the body's ability to meet a demand and with the body's ability to respond to therapeutic measures.

It is much easier to remedy an over-acid condition than an alkaline condition. In most chronic ailments, the tendency is to become increasingly alkaline, making recovery more difficult to achieve. Thus it is of prime importance in the treatment of illness, to have an accurate means of ascertaining the acid-alkaline balance in the body, and also to have an effective method of arriving at the type of diet which will correct any acid-alkaline imbalance.

In the past there have been various means for testing acidity and alkalinity, as for example in the blood, urine, saliva and faeces, but the development and use of the neuro-micrometer produced quite a surprise when it came to the measurement of this factor, for it was found that the conventional tests did not measure the all-important acidity or alkalinity in the *tissue fluids*, and it was found that a person can have a relative over-acidity of the blood and urine, and still have a systemic alkalinity Thus the "regular" tests have often been misleading, and by falsely signalling the need for an acid-neutralizing diet, have led to measures which only make the patient worse.

The reason why there can be a discrepancy in acidity measurements of the blood, saliva, urine, faeces and tissue fluids, is because the blood and the other factors change their acidity in relationship to each meal, and therefore fluctuate from meal to meal according to the effect of the food intake in creating more or less acid, whereas the acidity of tissues changes much more slowly. The acidity of tissues represents an average in the fluctuations between high and low levels of blood acidity.

MINERAL-VITAMIN FACTORS

Dr. Wilson's book *A New Slant to Diet* contains a review of the different minerals and vitamins now known to be necessary to health. His conclusions are, that vitamins do not act alone but only in and through the vital minerals; when vitamins are taken alone or with foods either refined or lacking in mineral salts, their beneficial effect is largely nullified. A well-balanced diet is essential if vitamins are to prove effective.

BULK FACTORS

will be discussed under "Toxins arising from insufficient elimination".

CELL POISONING, AND TOXICITY

We should keep in mind the main purpose of Dr. Wilson's research, which is to maximize the amount of neuro-electricity flow which can be evoked when needed, to meet demands. It is not necessary that a large quantity of neuro-electricity be flowing at all times—rather that it should be *able to flow*, and to be transmitted, when required by the needs of different portions of the body.

There are two main approaches to this undertaking, and both are of equal importance. The first is to supply in the diet, an adequate supply of foods high in life-energy units, as these are the foods that facilitate the production of functional energies in the body. This we have already discussed.

The second main approach is to avoid toxic conditions and other causes of cell poisoning, if for no other reason that poisoned cells cannot create or transmit the quantities of neuro-electricity required for health.

Naturopaths, some of the Chiropractors, and a few of the M.D.'s have preached for decades the necessity for keeping the body free of toxins and cell poisons, but in the absence of specific means for detecting and measuring objectively the presence of these adverse factors, the discussion was largely theoretical. Now with the neuro-micrometer, precise measurements are made, and the relationships between these factors and various states of health and illness have been delineated with scientific exactitude, and new and valuable conclusions have resulted.

Our presentation of this phase of Dr. Wilson's research is divided into three parts:

    I. Cell poisoning resulting from unbalanced intake or effect of foods.

    II. Toxins arising through inadequate digestion.

    III. Toxins resulting from insufficient elimination.

I.   CELL POISONING FROM UNBALANCED INTAKE OR EFFECT OF FOODS

Under this heading we find three types of conditions:

1. Over-acidity.

2. Over-alkalinity.

3. Toxins resulting from protein intake.

1. *Over-acidity*

This can come from any of three causes: deficiency of oxygen, fatigue, and wrong eating.

Considering first the deficiency of oxygen, it has been found that the neuro-micrometer's registering of "cell-oxidation" is a more accurate measure of oxygen consumption than is the basal metabolism test. The latter is only made when the body is at rest, and oxygen consumption is kept at a minimum, whereas cell-oxidation measurement is a true indication of the extent to which oxygen is reaching the cells of the body under conditions of more normal demand.

It has also been found that it is the leafy vegetables and fruit which increase the body's supply of oxygen in a form available for cell oxidation, and also for tissue-fluid oxidation, by means of which acids are burnt up in the body. In this way, excessive quantities of acid are neutralized.

Lactic acid, created as a result of fatigue, ordinarily causes only a temporary effect, about a day or two. The acids from wrong eating have a much longer effect, and are the prime cause of over-acidity in the body.

Tests with the neuro-micrometer show that an excessive acid condition can be created in the body within four hours by one unbalanced meal, and even if all further meals are properly balanced, from two to four days are required to bring the acid-alkaline balance down to normal in the healthy. In sick people where an over-acid condition has existed for some time, weeks or even months are required for its neutralization.

The acid-neutralizing foods are found to be primarily the leafy vegetables, and secondarily the fruits (except citrus) and meat proteins.

2. *Over-alkalinity*

In the past there has been no adequate method for detection of mild cases of over-alkalinity. Severe cases, which can be considered as having alkalosis, are hopelessly ill—it is then too late to save them. Therefore the value of having a means of detecting over-alkalinity of the body tissues in the early stages should be apparent.

Alkalinity involves chemical changes in the body, in which diet appears to exert an indirect influence. Two sets of factors are involved—the functional factors and the chemical factors.

(*a*) *Functional Factors* concern the extent to which "an over-alkalinity and an alkalosis are the result of functions gone wild, as are cancer, anaemia, leukaemia, dysentery, chorea, palsy, nervousness, etc."

"Back of functions going wild, is the body's power of response and what happens when it gets out from under proper control. It is one thing for the body to be able to respond to a demand; quite another for the response to stop when a demand has been met. An over-acidity, which is always present at the onset of an ailment, creates a demand to which the body responds by increasing the oxygen intake and stepping up the oxidation of tissue fluids to burn up the acids. While the over-acidity makes possible an accelerated nerve flow to the parts involved to co-ordinate their activities, and the stepped-up oxidative process serves a useful purpose, both are beneficial only so long as the demand exists. Beyond that point they become destructive. Thus it is just as important for the stepped-up activity to cease when the demand has been taken care of, as it is for the body to be able to meet the demand in the first place."

Unfortunately, pressure on the afferent nerves can prevent the "cease-firing" order from getting through when the battle has been won.

So the battle continues unabated to the point that not only the excess acidity has been overcome but a bad alkalinity has been created, and in this way the body can destroy itself.

(*b*) *Chemical Factors.* To quote again from Dr. Wilson: "Just why it is that the same mineral salts will form acids at one time and at another will form alkalines is difficult to explain except on the basis of hydrogen and hydroxyl ions. Briefly, ions come from the dissolution of mineral salts in the body. 'Whenever a mineral salt is dissolved in a solvent such as water' says Cook, 'the molecules break up into positively and negatively electrified particles called ions.' (Elements of Electrical Engineering.)"

"The positively charged particles are known as hydrogen ions; the negatively charged particles, hydroxyl ions. Thus it is obvious that the same mineral salts can form either acids or alkalines according to whether they become positively or negatively electrified (charged). Why they take on a positive charge at one time and form hydrogen ions, or take on a negative charge and form hydroxyl ions at another time, is controlled by . . . the state of a person's health, for the ions formed in a healthy body are always of the hydrogen type, and when unhealthy, the hydroxyl type. . . ." Without exceptions an over alkalinity will be found in chronic ailments, "and this fact is of great importance in treating the ailment."

(*c*) *The Correction of an Over-alkalinity*. Dr. Wilson shows that Chiropractic adjustments to remove pressure on afferent nerves is of importance in minimizing functional causes of over alkalinity; and that the chemical factors can best be overcome by the use of a diet which will build up the condition of the body cells. Such a diet would not be over-acid, rather it consists of well-balanced meals; for the alkalinity cannot be combatted directly—it yields only indirectly, as the health of the body cells is gradually improved.

3. *Toxins Resulting From Proteins*

Neuro-micrometer research shows that about 60% of the people have an over-acidity—the result of an excessive amount of carbon, while 80% of the people have a toxic condition—the result of an excessive amount of protein, or of undigested products of protein. Perhaps the reason for the greater prevalence of protein-poisoning as compared to acid-poisoning, is because the body has three ways in which to use or eliminate carbon, but only one way of handling nitrogen, the key element of proteins.

Carbon can: 1. Unite with other mineral salts to form different compounds, or
     2. be burnt up in the cell-oxidative process, or
     3. be eliminated as carbon dioxide.

Nitrogen can only unite with other minerals to form nitrates, with a toxin as the end effect.

Dr. Wilson is concerned not so much with the severe protein poisoning which has been termed "proteinosis"—an ailment manifesting bad headaches and general aching, as he is with the much more common condition of a less severe protein poisoning which he terms—"hyperproteina", in which there is an excessive quantity of mucus excreted by some portion of the mucus membrane. He finds that nature has no adequate provision for getting rid of waste protein products through the regular channels of elimination, and therefore they are largely thrown off in the form of mucus.

It is stated that all mucus ailments—sinusitis, colitis, appendicitis, bronchitis, pneumonia, colds and dozens of other ailments, are all the direct effects of hyperproteina, or proteinoids, as they are now called. Hyperacidity is usually present in these ailments, but Dr. Wilson finds that rather than being the direct cause of the formation of mucus, the acid-forming foods slow the digestive process to the point where proteins can not be completely split-up, leading to toxic protein products as will be described shortly in this review.

The correction of the widespread protein poisoning which neuro-micrometer measurements show is present in our population, lies not in eliminating proteins from the diet, for these are needed to maintain structural integrity of the cells, as previously stated. Nor does it lie in the use of vegetable proteins rather than animal proteins. It is found, for example, that thin people digest vegetable proteins even less readily than animal proteins. The solution lies rather in the selection of *food combinations* which will favour the more complete digestion of proteins so that toxic products will not be formed. This subject is included in our next topic of discussion.

II. TOXINS RESULTING FROM INADEQUATE DIGESTION

These arise principally from incomplete digestion of proteins. "Proteins as such cannot be utilized by the body. It is only as they are split up into amino acids that they can be reformed into the proteins of which the human body is made. The digestion of proteins is technically known as 'splitting up' the protein molecule, which goes through seven processes," listed in detail in Dr. Wilson's book. The fourth in the chain of processes produces peptones, the first stage at which the protein products can be *absorbed* into the blood stream. However, *absorption* and *assimilation* are two different matters, as applied to proteins. If the peptones go through the final three stages of the digestive process, they become amino acids and can be adequately assimilated into the body cells. Too often, however, the digestion of proteins is sluggish, with the result that a variable proportion of protein intake is absorbed into the blood at the peptone stage.

The effect of absorption of peptones, is to clog the inter-cellular spaces with mucus, thus impeding the circulation of tissue fluids around the cells. It is this fluid circulation which carries the vital mineral salts to the cells of the body, hence the result of incomplete digestion of proteins is to lessen the quantity of mineral salts that can reach the body cells; hence the latter deteriorate in condition. This is one of Dr. Wilson's major discoveries. (Many have considered mucus to be largely carbon, but Dr. Wilson's chemical analysis has instead shown mucus to be largely protein.) If the body's power of response (to meet demands) is to be built up in sick people in whom this factor is low, the condition of the individual cells of the body must be improved, and freeing the circulation of inter-cellular waste is one of the main considerations that need attention towards this end.

Having discussed the effect, the next point to consider, is the cause of the presence of incompletely-digested protein products in the blood stream. To say that the condition is brought on by too much meat is both right and wrong; "right concerning too much protein being in the fluids of the body, but wrong concerning the condition being caused by an over-consumption of meat." Too much meat can be eaten, it is true; but more often it is not meat in particular or the consumption of proteins in general that is at fault. It is the process of digestion. If digestion is complete and the molecules of protein are split-up into amino acids, there will be no mucus. The amino acids would then be assimilated. Moreover, amino acids do not cause mucus, "for mucus is partially digested protein, while amino acids are completely digested proteins."

Dr. Wilson's studies point the way towards eliminating the production of protein toxins, through improving the digestion of proteins. This improvement is brought about partly through changes in the relative proportions of the different classifications of foods in the diet, and partly through selection of combinations of foods to be eaten at the same meal, to the end that digestion will be improved rather than impeded.

The digestive secretions required for assimilation of proteins are different from the secretions required for the proper digestion of carbon foods—sweets, starches and fats.

The secretions for the digestion of either type of food, flow in response to a demand. When both types of foods are eaten at the same meal, the demands are mixed and the response in secretion flow is impaired as a result. The effect is that either or both of these two main types of food will be incompletely digested. If too much of the carbon foods are eaten, or if they are incompletely digested, then an excessive quantity of acid is created. If too much of the protein foods are eaten, or if they are incompletely digested, then toxic waste material and mucus is created. Each type of food interferes with the digestion of the other type of food. For example, hydrochloric acid is required for the digestion of proteins, but when starches, sweets or fats are taken at the same time, then less hydrochloric acid flows, so the digestion of proteins becomes incomplete.

To facilitate the digestion of proteins, the other foods to be eaten at the same meal should be leafy vegetables cooked and raw, and not more than one slice of bread. Fruits should be eaten between meals or as separate meals, since when taken with proteins they interfere with protein assimilation. Carbon foods—the starches, sweets and fats, should definitely not be eaten at the same meal with proteins, for reasons already discussed.

The proportion which each class of food should bear to the total food intake, differs according to the physical type. Dr. Wilson in his book, delineates the thin nervous type, and two types of fat people—the hydrated and hydrocarbonated. To learn how to differentiate between those two fat types, his book should be read.

The thin person he designated as the dehydrated type. These people have an intolerance to carbon foods, yet they often crave them and eat them in abundance. He advocates that dehydrated people include ample protein in their diet, a minimum of carbon foods and plenty of leafy vegetables cooked and raw. Among the protein foods, he finds that meat, cheese, eggs and avocado are best; nut, dried legume and whole-wheat proteins are not so well handled by this type. Salt intake should be increased, to aid in the absorption of water into the tissues.

For the hydrated fat type, the diet advocated is similar to that for the dehydrated, except that not as much salt is included.

The hydrocarbon fat-type has an intolerance for proteins, yet usually craves them. Hydrocarbon people seem unable to digest proteins without creating an excess of harmful bacteria. Therefore it is advocated that hydrocarbon people eat protein foods very sparingly, and concentrate almost entirely on vegetables, grains, and fruit, with the fruit eaten separately. The protein intake should be from such foods as legumes, cottage cheese, milk, eggs, etc., rather than from meat.

Additional details on diets for these different physical types, and modifications of the diets in various disease conditions, are given in Dr. Wilson's book *A New Slant To Diet*.

III. Toxins resulting from Insufficient Elimination

The importance of good elimination is universally recognized. The fact that very few people achieve good elimination, even when evacuations *appear* to be satisfactory, is not so widely known. The facts in the matter are readily proven both by autopsies and by colon X-rays. Dr. Wilson quotes from *The Royal Road to Health*, written by Dr. Charles A. Tyrrell, M.D.

> "I found the fountain of premature old age and death, for, surprising as it may seem, out of the 284 autopsies held of late on the colon (these cases representing nearly all the diseases known to our climate) only twenty-eight colons were found to be free from hardened, adhered matter, and in their normal state. . . . In many of (the autopsies) the colon was distended to double its normal size throughout its whole length, with a small hole through the centre, and as far as could be learned, these cases had regular evacuations of the bowels each day. . . ."

Of the many X-ray pictures taken of colons in Dr. Wilson's clinic in Utah, only one out of fifty, on the average, was normal. Even the fact of frequent bowel movements does not assure good elimination, since too often when the bowels move two to three times a day, they do so because of fermentation which develops an acidity. The acidity irritates the bowels and causes them to evacuate frequently. It all takes place because of the retention of excreta too long in the cecum (a portion of the colon), as a result of the cecum being enlarged.

Poor elimination means a rapid build-up for toxins in the body, which in turn poisons the cells and prevents them from functioning normally. Inasmuch as very few of us have good elimination, likewise very few people indeed enjoy the really excellent state of health which should be theirs.

To illustrate more clearly the role of toxic products in health and survival, we need only turn to the experiments of Dr. Alexis Carrel, who kept the heart tissue of a chicken alive in the laboratory by periodically washing out the poisons produced by its own life processes. Repeatedly it became senile, and about to die, but was at once rejuvenated by a thorough washing out.

"It will be seen, therefore, how extremely important it is to reduce our daily dose of poisons and to eliminate as thoroughly and promptly as possible those which are unavoidably produced."

To induce good elimination, Dr. Wilson advocates:

Chiropractic adjustments to relieve pressure on the nerves to the bowels, so that the bowels can function more normally.

Proper diet, by which he means a diet of natural foods having a high proportion of bulk factors. It is the highly refined foods that are the most difficult to eliminate from the system.

Colonics for a short period of time, to give the bowels a good cleansing out, but not to be continued for permanent relief.

Plenty of fluid intake, including the consumption of two to six glasses of hot water each morning upon arising, as taught by Dr. Leo Spears.

After following this programme, the neuro-micrometer always registers a substantial reduction in body toxins, showing that an important step towards better health has been taken.

THE LEAFY VEGETABLES

Among the most important conclusions to emerge from the neuro-micrometer research, are the findings regarding the great benefits obtainable from leafy vegetables. While it has been known for some time that these were "protective foods", there has previously been no way to determine just what their effect was upon the body's functioning. Through electronic measurements, it is now known that it is the leafy vegetables that have the highest life-energy ratings, also that these are the only foods which satisfactorily balance a protein-content meal in such a manner that the body can assimilate protein without reducing one's store of functional and reserve energies. In fact, if the Wilson instructions are followed, these energies can be increased, which has the all-important effect of strengthening and building up the body's "power of response"—the power to meet demands; the power to recover from illness.

The tables of life-energy units compiled from the Hemoelectrometer measurements of the different foods, show that the leaves of vegetables have a much higher rating than do the stalks, pods, legumes or root vegetables.—"Only one answer as to why this is true has been found, and we need not look for another answer, for it is quite satisfactory. It is the action the sun rays have on the leaves of vegetables. Just why or how the sun rays have or bring about such a beneficial action is a secret many scientists have sought in vain to solve."

Dr. Wilson advocates that when vegetables are eaten which have lower life-energy values, not more than one of these should be eaten at a meal, and it should be balanced by at least two vegetables having the higher life-energy factors, and one of those two vegetables should be raw. That is because dead elements do not sustain life in the way that live elements do; cooking reduces the life-energy ratings of foods.

It is further stated that the best vegetables from the standpoint of their content and life-energy ratings, are: red beet leaves, celery stalk and leaves, mustard greens, kale, lettuce, endive, turnip greens and carrot tops, alfalfa and water cress.

To tell people what they should eat for health, and to get them to eat accordingly, have turned out to be two different matters. Many people cannot be induced to eat the quantities of leafy vegetables that are found to be necessary for the restoration or maintenance of good health. As an alternative measure, it was determined that the preparation of tablets made of powdered concentrates from the dehydrated raw vegetable leaves provides a satisfactory substitute. Details regarding these concentrates, the recommended dosage, and sources of supply, are contained in Dr. Wilson's book.

CO-OPERATIVE HEALING

(*Developed by L. E. Eeman*)

INDEX

CO-OPERATIVE HEALING

1. THE CENTRAL THEME

THIS term refers to therapeutic effects obtained by applying wire circuits to connect one part of a person's body with another part, or to connect an area of one person's body with an area of another person's body. L. E. Eeman, in his exhaustive research into this type of circuit as detailed in his book *Co-Operative Healing*, utilized these connections in pairs so that there would be a complete electrical circuit.

The basis of the flow effects obtained by the use of these circuits, was the discovery that human bodies give evidence of being bi-polar, along three axes—

Head to feet
Right side to left side
Back to front

When polar opposites are connected together, whether within one body or between two or more bodies, the effects noted include greater relaxation, recovery from fatigue and disease, greater capacity for work, better health in general. Circuits connecting polar opposites are termed "Relaxation Circuits".

When polar similars are connected together, the effects noted are the creation of tension, and the opposites of all the advantages obtained from Relaxation Circuits. Circuits connecting polar similars are termed "Tension Circuits".

The use of these circuits have been found to affect not only functional physical and mental conditions but also organic conditions.

In people born right-handed, the head and right side have been found to have the same polarity, which by convention has been termed positive. In these individuals, the base of the spine, and the left hand share the same polarity, opposite to that of the head and right side, and is therefore termed negative. Polarities are reversed in those born left-handed.

Eeman and co-workers performed thousands of carefully documented experiments under controlled conditions over a period of years, establishing the effects of a wide variety of these circuits, and ruling out the possibility of effects through suggestion or other extraneous influences. For more detail, see his book *Co-Operative Healing*, published in London, 1947 by Frederick Muller, Ltd.

2. ONE PERSON IN CIRCUIT

The diagram (Fig. 1) shows the circuit. Two metal handles each connect through approximately 10 ft. of insulated copper wire, to copper screen mats about a foot square. The individual lies on his back with the copper mats under his head and base of spine.

For right-handed individuals, a relaxation circuit is formed if the left hand holds the handle connecting to the mat under the head, and the right hand holds

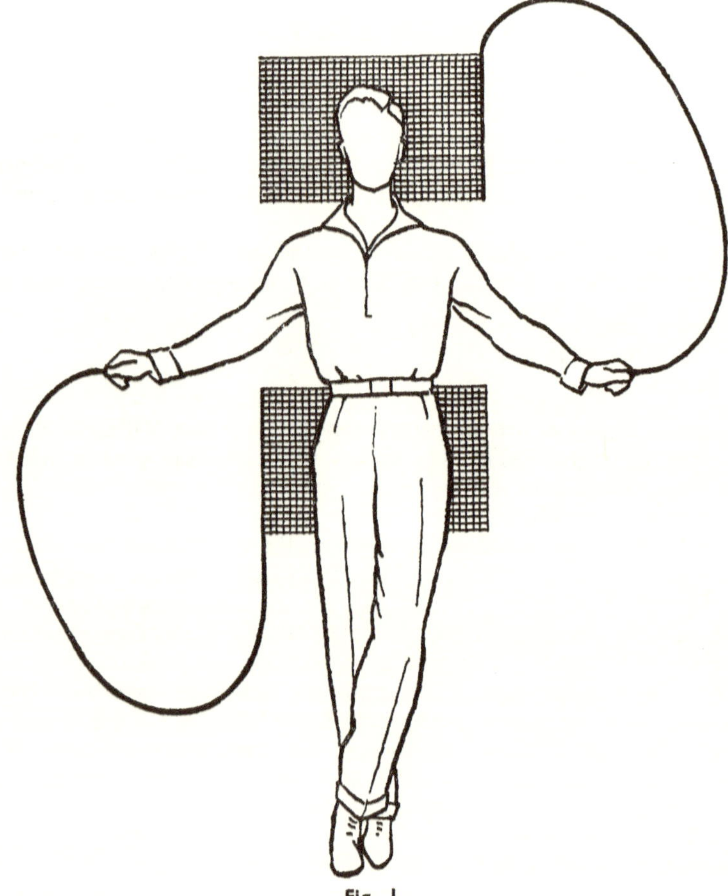

Fig. 1

One subject in relaxation circuit, showing copper gauze mats and wire connections.

the handle connected with the mat under the bottom of spine. Average duration of exposure is half an hour.

Likewise for right-handed individuals, if the handles are switched so that the left hand holds the handle connecting to the mat under the bottom of the spine, and the right hand holds the handle connecting to the mat under the head, a tension circuit is formed. These effects are noted even if the individual is not aware of the specific connections or if the connections are changed out of range of his vision and without his knowledge. One of the reasons for 10-ft. length of connection wires, is to permit changes of connection without the subject's knowledge, in order to test the effect of suggestion. In almost all instances, the reactions are those appropriate to the circuit rather than to the suggestion given, if the suggestion is contrary to the effect of the circuit.

Many case reports covering various types of ailments are detailed in the book *Co-Operative Healing*, and will not be repeated here. Among the ailments relieved by these circuits are listed: mental, nervous, circulatory, respiratory, digestive and eliminative disorders; headaches, high blood-pressure, rheumatism, lumbago, sciatica and many other conditions. Insomnia appeared particularly to be relieved.

Eeman states:

"The relaxation circuit almost invariably produces a progressive sense of muscular relaxation, warmth, well-being and drowsiness, often culminating in sleep, slower and stronger pulse, slower and fuller respiration, with more complete deflation, progressively longer pauses between deflations and inflations, and a lowering of the pitch of the voice."

The Tension circuit reverses the above reactions and eventually leads to varying degrees of tension, restlessness and discomfort.

It was found that direct contact between the copper mats and the individual was not necessary in order to produce the effects noted; clothing and even cushions did not act as barriers.

3. TWO PERSONS IN CIRCUIT

The effects of both Relaxation and Tension circuits are usually increased if two persons are connected with each other, using a separate pair of copper mats for each person.

The normal Relaxation circuit is formed when the head of each person is connected to the left hand of the other person, and the bottom of the spine of each person is connected to the right hand of the other person. See Fig. 2.

It was found that sex made no difference in polarity—the same circuit being applicable to two men, two women, or one woman and one man. However, Left and Right-handedness does make a difference in polarity. When a right-handed person is placed in circuit with a left-handed person, then the Relaxation circuit is formed when the head of each person connects with the right hand of the other person, and the bottom of spine of each person connects with the left hand of the other person. Also, the left-handed person should lie in the opposite direction from the right-handed one.

Reports from the experiments indicate the rather curious fact that when a strong or well person is placed in circuit with a weak or ill person, both experience beneficial effects from the experience. Sleep for a half hour or more will frequently visit both subjects.

Another interesting fact that emerged from the tests, is that a Tension circuit tends to produce discord and antagonism between the two participating subjects, and this antagonism is then dissipated if a Relaxation circuit is used.

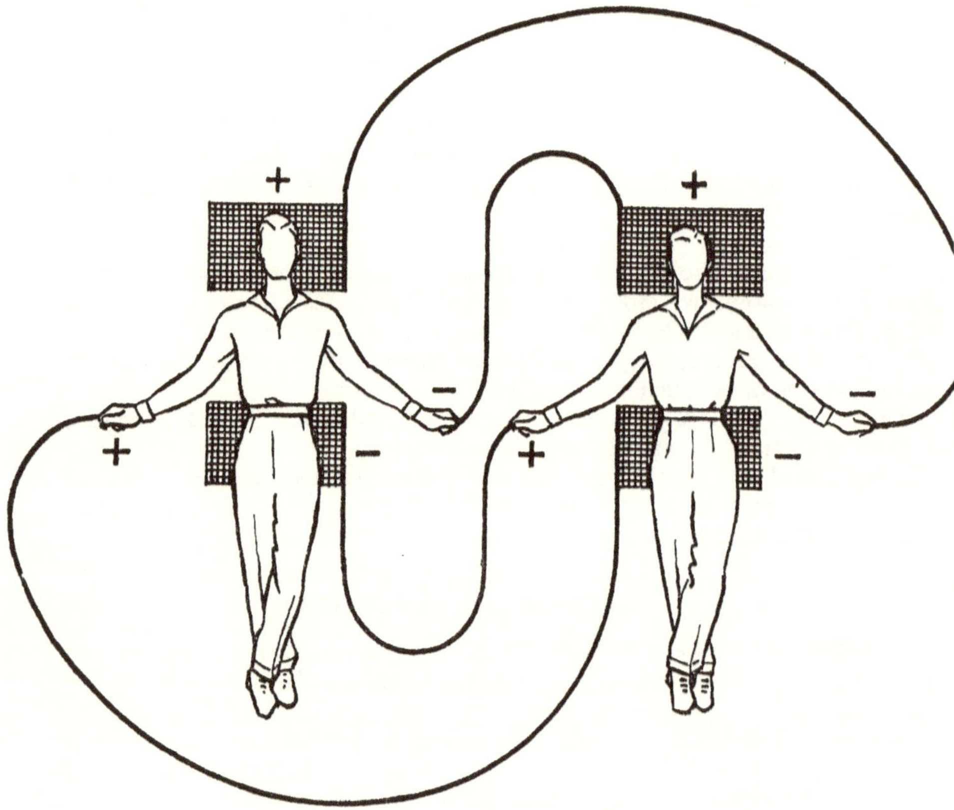

Fig. 2

Two subjects in closed relaxation circuit. Each subject with left hand (−) to head (+) and right hand (+) to spine (−) of other subject.

4. THREE OR MORE PERSONS IN CIRCUIT

The same principles were found to apply with three or more persons in circuit —in fact the number of persons could be increased indefinitely. However, with the addition of more individuals in the circuit, additional phenomena appeared; some individuals turned out to be stronger "emittors", other individuals acted more as "receivers", while others seemed to have as the primary characteristic,

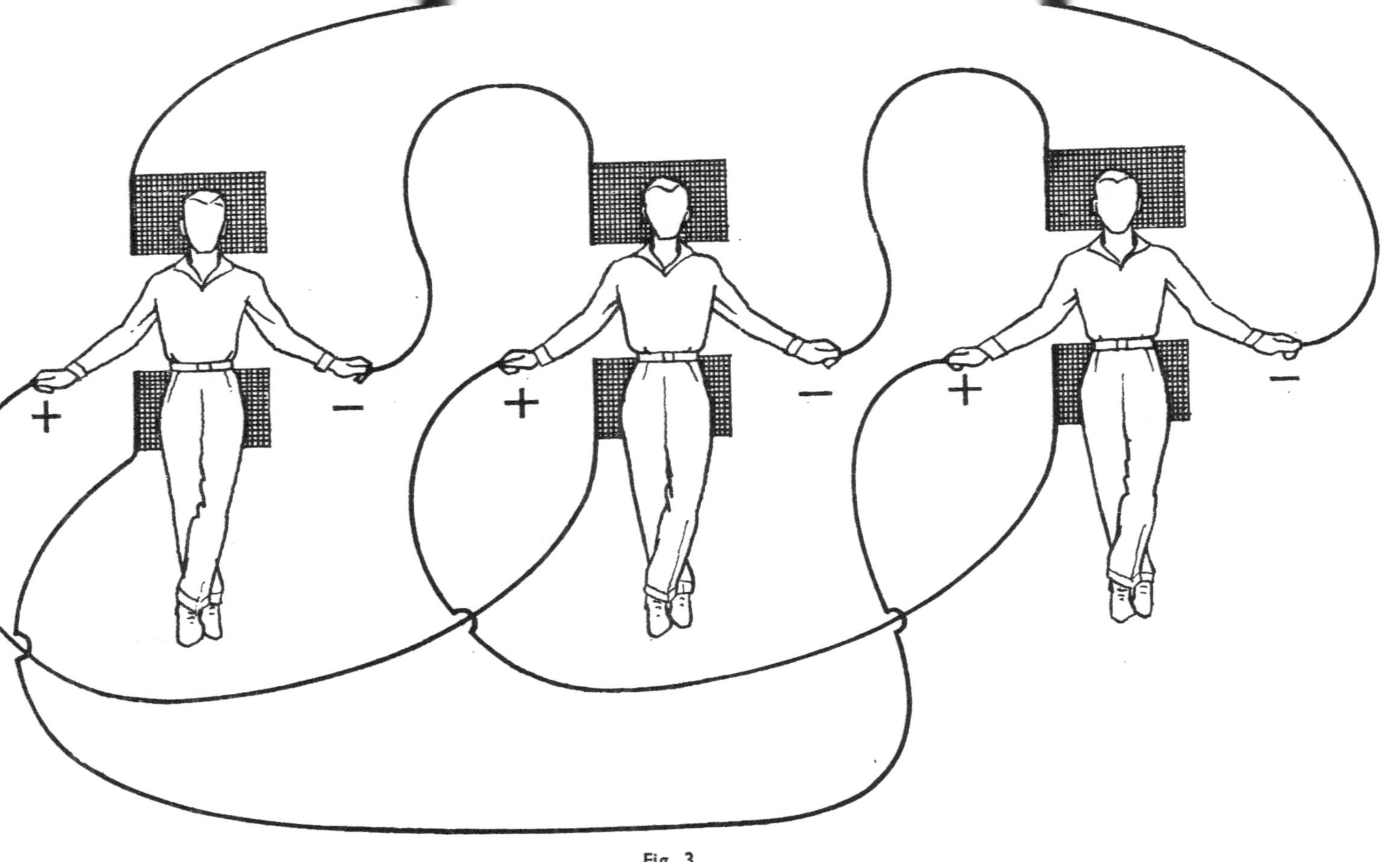

Fig. 3

Three subjects in serial relaxation circuit, Left and Right of (1) to Head and Spine of (2) and so on.

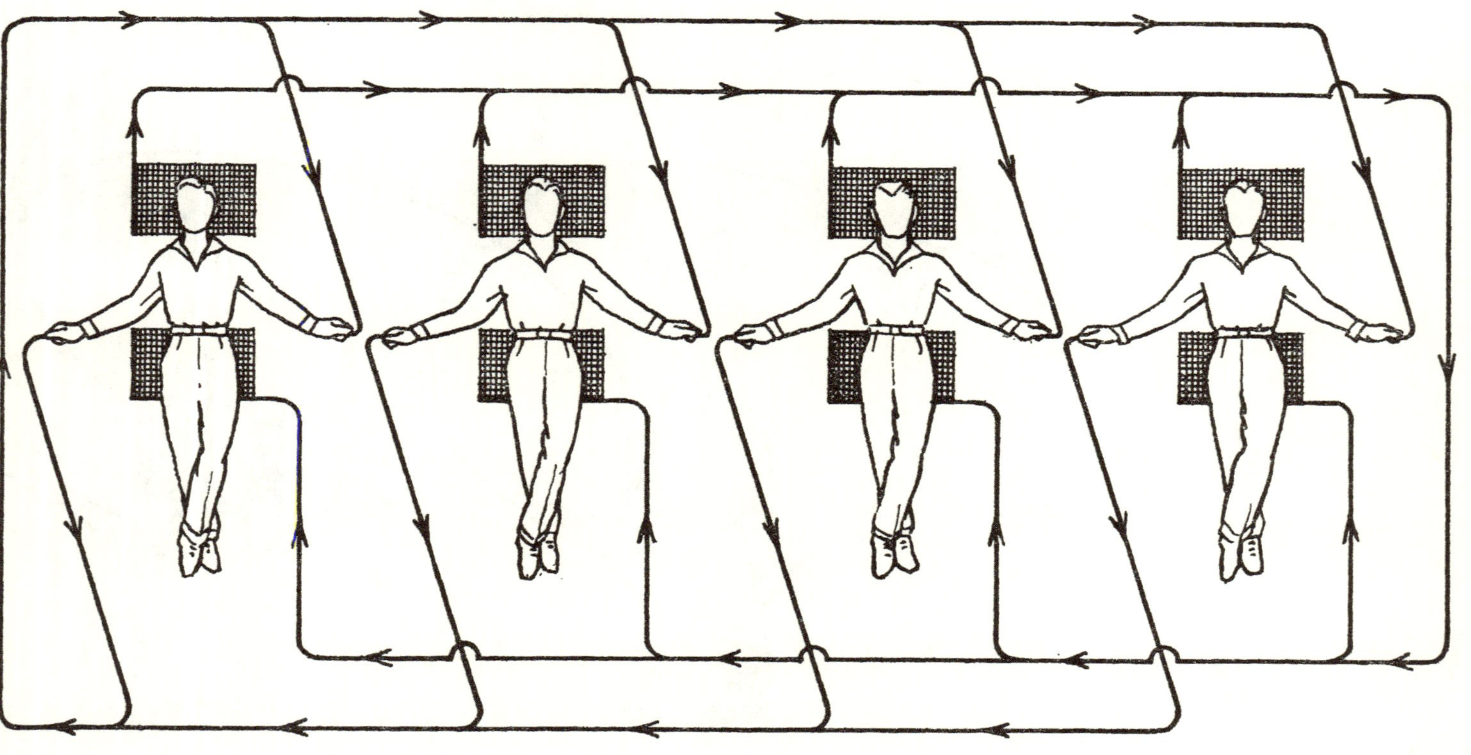

Fig. 4

All Heads to all Spines and all Rights to all Lefts, in pure parallelism.

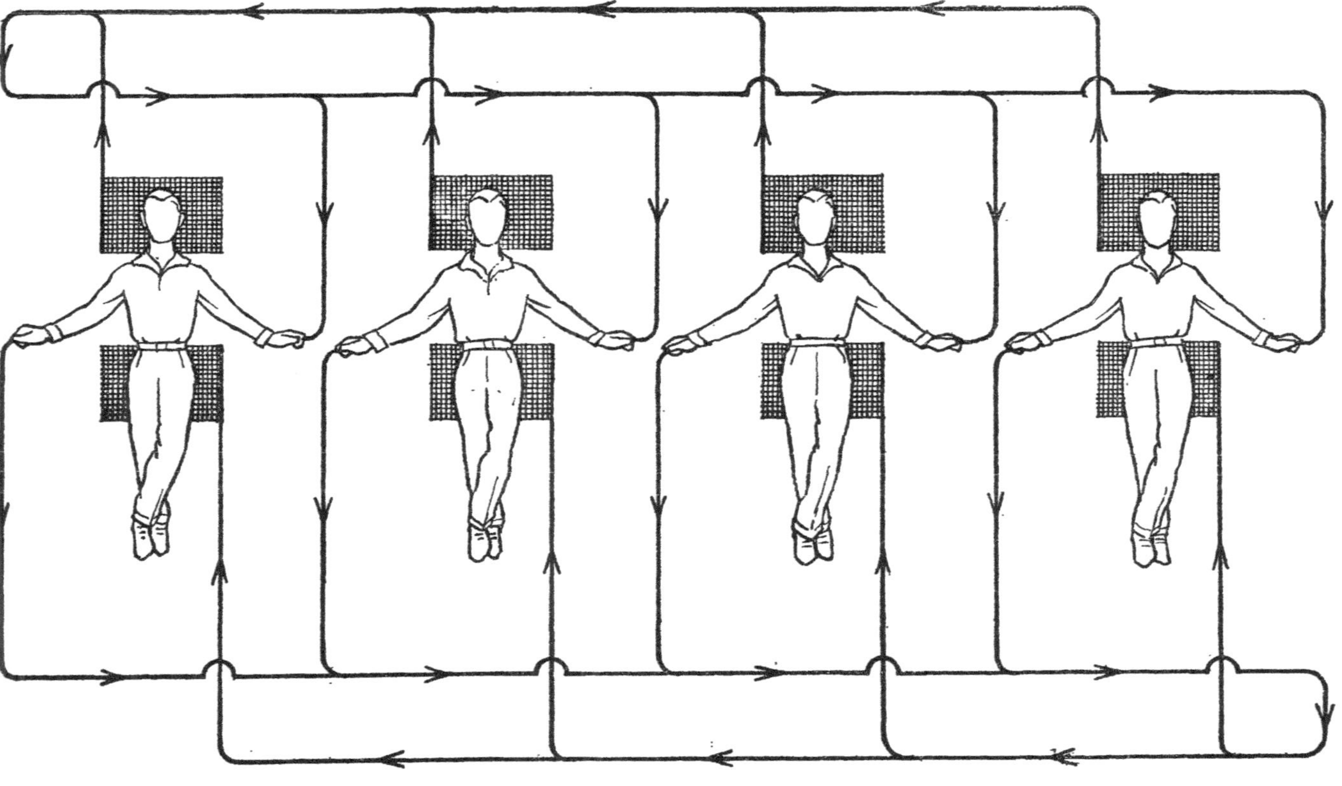

Fig. 5

All Heads to all Lefts and all Rights to all Spines, in pure parallelism.

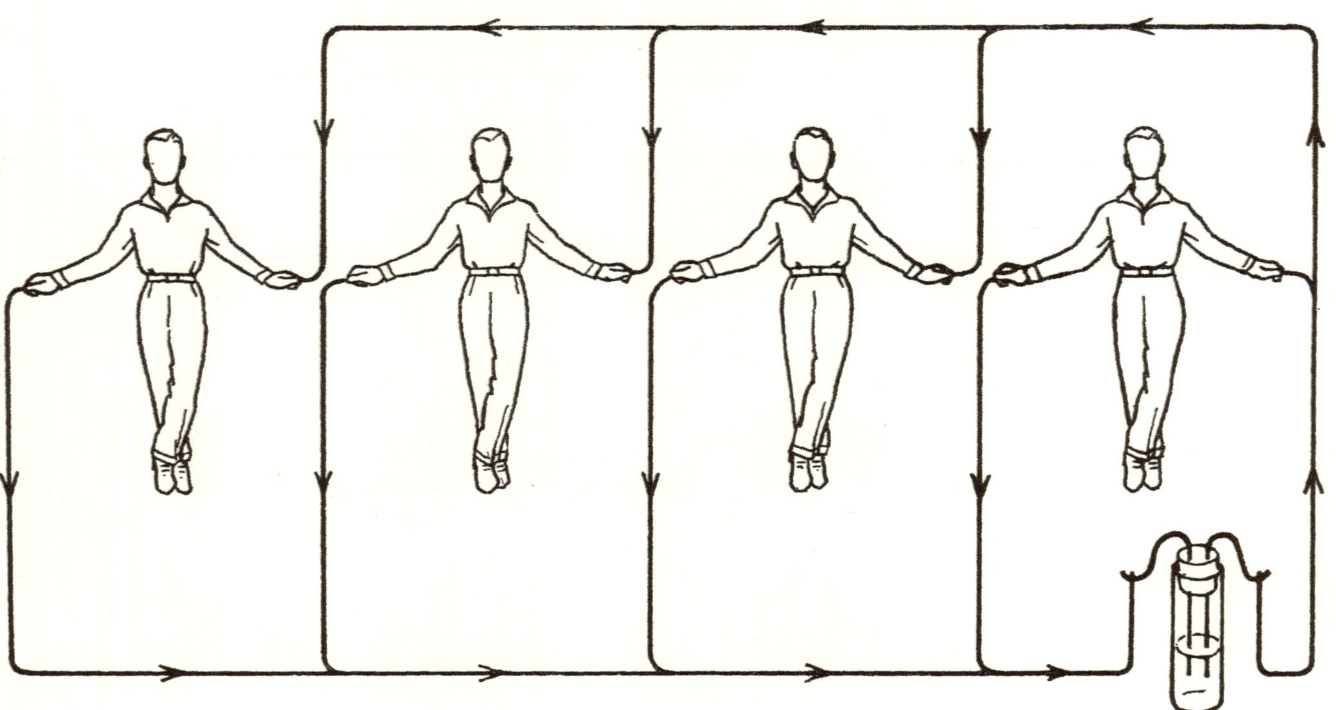

Fig. 6

Relaxation circuit using hands only in parallel, with drug in series.

that of being "conductors". For a detailed account of these characteristics, refer to Eeman's book *Co-Operative Healing*, Chapter IV.

Fig. 3 shows a Relaxation circuit between three individuals; the same circuit can be expanded to include any desired number. This is termed a "series" type of circuit.

For multiple use (three or more persons in circuit at one time), two principal circuits emerged as the best ones to use—diagrammed in Figs. 4, 5 and 6.

In Fig. 4, all the heads connect together, all spines connect together, and those then are connected with each other. All right hands connect together, all left hands connect together, and those two sets are likewise connected with each other.

In Fig. 5, all heads and all left hands connect together, and all spines and all right hands connect together.

5. CONCLUSIONS

It would appear from the Eeman experiments, that human beings emit, emanate or radiate a particular kind of current or impulse, which can be channelled with wire connections and used to produce either healing effects or adverse effects, depending upon polarity.

6. DRUG TESTS

One of the most interesting aspects of the Eeman research, was the series of tests in which a Relaxation circuit was established for one or several persons, and the circuit was broken by test vials placed in the circuit one at a time, each vial containing two electrodes immersed in a drug or medicinal solution. The vials were inserted out of range of the vision of the subjects, and they did not at any time have any knowledge of the names or characteristics of the test substances. Yet within a short time after a substance had been inserted, the subjects experienced physiological changes and symptoms, which almost always were representative of the effects engendered by the drug or substance used in the test. If the subjects were kept in circuit with a substance for too long a time, there were instances where the symptoms or physiological effects became very acute and severe.

Some of the tests were run with coded vials, so that the operator would not know what substance was in use at any given time, and so could not have had telepathic influence over the minds of the subjects. These tests, conducted to rule out the possibility of telepathic effect, proved just as positive in results as the tests conducted when the operator (though not the subjects) knew what substances were being tested.

Fig. 6 shows a Relaxation circuit for drug tests.

The implications of these tests tend to confirm the hypothesis, previously stated by others, that all substances radiate and that these radiations can be carried on wires.

CHIROPRACTIC

CHIROPRACTIC

PRINCIPLES OF CHIROPRACTIC

In brief, the principles of chiropractic are:

1. That the maintenance of normal function and a state of health in the various organs and systems of the human body, depend upon the free, unimpeded flow of nerve energy to and from the various organs and tissues of the body.

2. The nerves, that control the function and state of health of the different organs and tissues, follow a path leading from the brain down through the interior of the spinal column, and branch out between the different spinal segments (vertebrae) to their respective ultimate destinations in the body.

3. The nerves as they emerge from the spinal column between the different spinal segments, are subject to abnormal pressures if those vertebrae are not in proper alignment. These pressures result from the fact that the space for the nerves and allied vessels is ordinarily just sufficient in a well-aligned spine; it will be seen from an examination of the mechanical relationships involved, that any mal-alignment of the spine has the effect of diminishing the space available to the nerves and their associated structures, therefore any mal-alignment puts abnormal pressures upon the nerves and their blood vessels and lymphatic vessels.

4. By virtue of the fact that we have a flexible spine, the individual spinal segments must be capable of a certain amount of movement relative to each other. Ordinarily the vertebrae automatically return to their normal alignment after movements of limited scope have taken place. However, various circumstances connected with the use and abuse of the body in everyday living, may at times result in an excessive movement of individual vertebrae with relation to each other, beyond the extent that can be corrected by automatic realignment. The result is mal-alignment or mal-positioning of one or more of the spinal segments. The Chiropractic term for this mal-alignment is "subluxation".

5. A subluxation is not as extreme an abnormality as a dislocation; however, it occurs much more frequently than do dislocations, and constitutes a shift in or a departure from the normal relationship of the surfaces of adjacent vertebrae in the spine.

6. Pressure upon the nerves resulting from spinal sublucations, produces what chiropractors term "impingement", which reduces the ability of the nerves to function. That is, their ability to transmit impulses is impaired, therefore the organs and body tissues served by those nerves are forced to operate under a handicap.

7. Organs and body tissues supplied by impinged nerves suffer reduced function, abnormal function, and are peculiarly susceptible to invasion of bacteria, virus, and other disease-producing agents. Thus subluxations resulting in nerve impingement can be either a direct "exciting" cause or an indirect "predisposing" cause.

8. The science of Chiropractic teaches various techniques of adjustment by which proper alignment of the spine can be restored, thus removing impingement from the nerves, permitting restoration of normal function for nerves whose functioning has been impaired by impingement. With the renewal of adequate nerve function, the condition and functioning of organs and body tissues supplied by those nerves usually improves to a marked extent, which shows why spinal adjustments quite frequently result in cures of ailments which have resisted other forms of treatment.

Having briefly listed the principles of Chiropractic, we shall now discuss these principles, with particular attention to:

No. 1. that the organs and body tissues are controlled by the nerves,
No. 2. that subluxations reduce the functioning of the nerves passing through the subluxated area,
No. 7. that impingement of nerves due to subluxations, results in abnormal function, inadequate function, and disease processes in the organs and body tissues served by impinged nerves.

1. THE DEPENDENCE OF ORGAN STRUCTURE AND ORGAN FUNCTION UPON THE MAINTENANCE OF FREE FLOW OF NERVE ENERGY

This is shown by the following facts:

(*a*) Nerves connect all parts of the body, for co-ordinated functioning.

(*b*) Organ function is controlled by outgoing impulses, and nearly all the outgoing impulses from the central nervous system are the result of the central nervous system's reaction to incoming impulses—as, for example, the flow of digestive juices that occurs after one sees food.

(*c*) The functional activities of all parts of the body are dependent upon the amount of nerve energy sent them, just as muscle effort in lifting is dependent upon nerve control impulses arising out of our judgment of weight of the object to be lifted.

(*d*) Temperature of the different parts of the body is determined by nerve control—if one severs or puts pressure upon the nerves to a limb, the limb grows cold. The same applies for any internal organ also.

(*e*) Nerves control the diameter of blood vessels. Diminution of blood supply to any part, reduces all vital activities.

(*f*) Besides the vaso-motor nerve functions just mentioned, there are other nerves which control the degree of metabolism (heat production through combustion of oxygen) in the various tissues. With insufficient metabolism, temperature becomes subnormal—in toxic conditions, internal irritations are set up which induce the nerves to increase the metabolic activity—hence fever.

(*g*) When there is no impediment to the free flow of nerve energies, fever is limited and beneficial, but with interference to the normal control channels, fever can become high enough to endanger life.

(*h*) Experiments with nerve severance in animals show devastating results on circulation, temperature, metabolism and nutrition of the parts ordinarily controlled by the nerves that have been severed.

(*i*) The flow of all internal secretions is determined by nerve control. Each of these secretions play a major part in vital functions of the body.

(*j*) All excretory functions are under nerve control. Adequate excretions of waste products from the colon, urinary bladder and skin are of prime importance in the maintenance of health.

(*k*) All operation of the sense organs (organs of sight, hearing, taste, touch and smell) are dependent upon nerve functioning.

(*l*) The activity rate of all internal organs, such as the lungs, heart, liver, spleen, stomach, colon, kidneys, etc. is determined by nerve control.

The above facts should be ample evidence of the vital importance of adequate nerve-energy flow in the maintenance of health.

2. THE SIGNIFICANCE OF THE SPINAL COLUMN, WITH REFERENCE TO PROPER FUNCTIONING OF THE NERVOUS SYSTEM

The control nerves come down inside the spinal column and branch off between the different spinal segments.

This is illustrated in diagram (1) which shows the different spinal segments, the nerves branching off from the spinal cord and emerging between the different segments, and pictorial representations of the different organs supplied and controlled by those nerves.

3. ANY MAL-ALIGNMENT OF THE SPINE, PLACES ABNORMAL PRESSURE UPON THE NERVES EMERGING BETWEEN THE SPINAL SEGMENTS OF THE MAL-ALIGNED PORTION

Each spinal segment has a hole in the centre, through which the spinal cord passes. From the spinal cord, nerves to the various internal organs emerge between the different spinal segments, through other holes formed by grooves in the spinal segments.

It is a well-known fact that nature permits no structure in the body to be unoccupied. The holes in the spinal segments are barely large enough to contain the nerves and vessels that pass through them, like cord inside of tight-fitting beads.

Likewise, the space allotted to the nerves emerging between the segments is

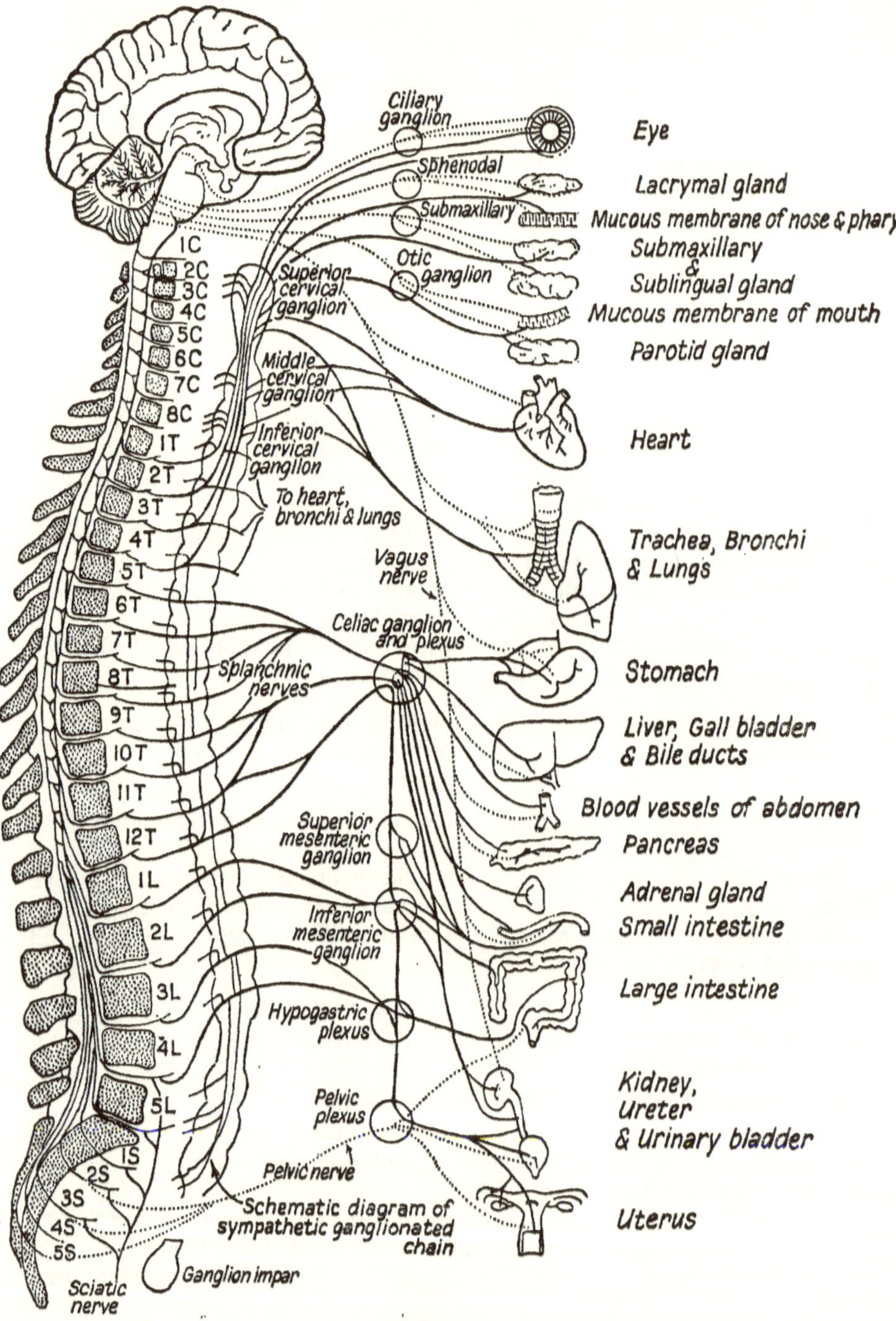

Fig. I

Autonomic nervous system, showing innervation of various organs.

178

just sufficient to house those nerves without interference with their function, when the spine is normally aligned.

Mal-alignment squeezes edges of some vertebrae closer together—see diagrams (2) and (3)—thus they encroach into the space needed for the nerves emerging between those vertebrae—see diagrams (4) and (5).

4. WHY DOES A SPINE BECOME MAL-ALIGNED?

There is a certain amount of motion that is normal, between any two adjacent vertebrae. Were the vertebrae locked in position, the spine would be rigid instead of flexible, and humans would be unable to make many of the movements which we find convenient in everyday living. As long as the normal motion between vertebrae is not exceeded, the bones may easily return to the normal resting position, without producing any abnormal effect upon any associated structures.

But the fact that some degree of movement between individual vertebra is possible and frequently occurs, also means that greater degrees of movement may take place, depending upon the force applied. When this movement exceeds certain definite limits, there exists the danger that the vertebrae will be unable to return to their normal positions.

Very slight movement is implied here, since motion of vertebrae amounting to $\frac{1}{8}$ in. will produce pressure upon the nerves passing between those vertebrae, sufficient to impair the conduction of impulses to the organs for which they are destined, with consequent adverse effects in the organs or parts supplied by the nerves affected by the pressure.

Vertebrae are held in position by strong bands of ligaments and muscles on each side. These protective structures exert a constant pull from both sides, and the balance between these pulls is intended to keep the spinal column aligned and erect. However, if muscle tone becomes poor, a sudden strain on the vertebrae may not meet with the proper resistance, and excessive movement between vertebrae may occur beyond the extent of correction by the automatic alignment processes. Such a sudden strain may occur from a fall or stumble, from a blow on the spine, or the thrust exerted from improper lifting of a heavy object with the spine at an angle, or from internal causes which will be discussed later.

It is interesting to note that when the spine is in the horizontal position—the position customary for most animals, the vertebrae lock with one another in perfect alignment—while in man the spine is held erect with an unbalanced weight at the top (the head, whose centre of gravity is located well forward from the spine) and a constantly varying system of forces and pulls is necessary for the maintenance of the erect position.

As an engineer, William Jay Dana, B.S., stated: "... The (human) spine is used as a column (erect position) while it is *designed* to serve as a beam (horizontal position). As a column it is far from ideal." Mr. Dana then goes into a

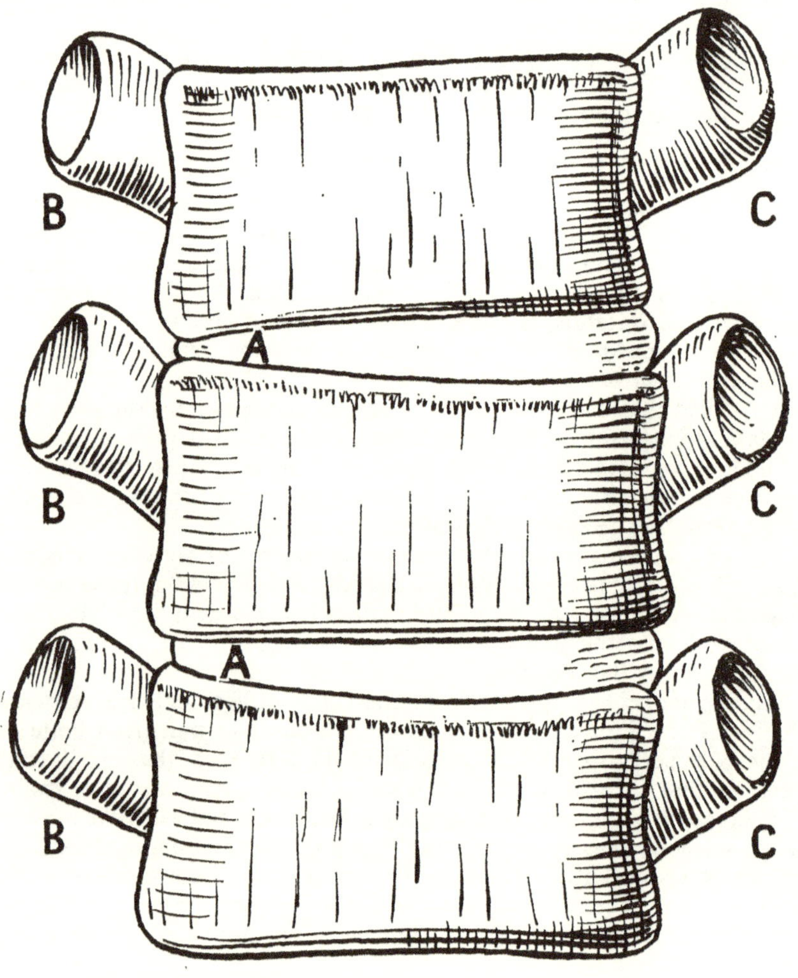

Fig. 2

Scoliotic Subluxations. Involve at least three vertebrae and are due to a thinning of the lateral aspect of the intervertebral discs (A). As a result, the transverse process on the contracted side (B) become more closely approximated, while those on the opposite side (C) become more widely separated. This results in a diminution of the vertical diameter of the intervertebral foramina on the compressed side.

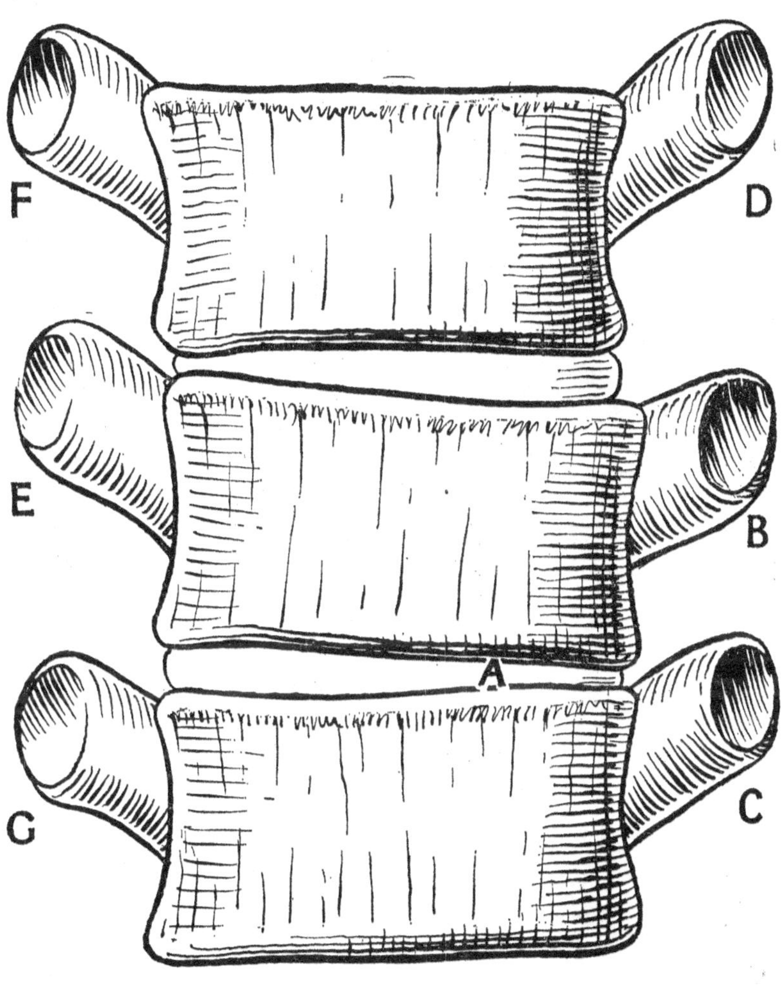

Fig. 3

Left inferior subluxation. There is a thinning of the left portion of the intervertebral disc (A) upon which the sub-luxated vertebra rests. The left transverse process (B) more closely approximates that of the vertebra below (C) and becomes more widely separated from that of the vertebra above (D). The right transverse process (E) more closely approximates that of the vertebra above (F) and becomes more widely separated from that of the vertebra below (G). As a result, there is a narrowing of the left intervertebral foramen below the affected vertebra and also a narrowing of the right intervertebral foramen above.

181

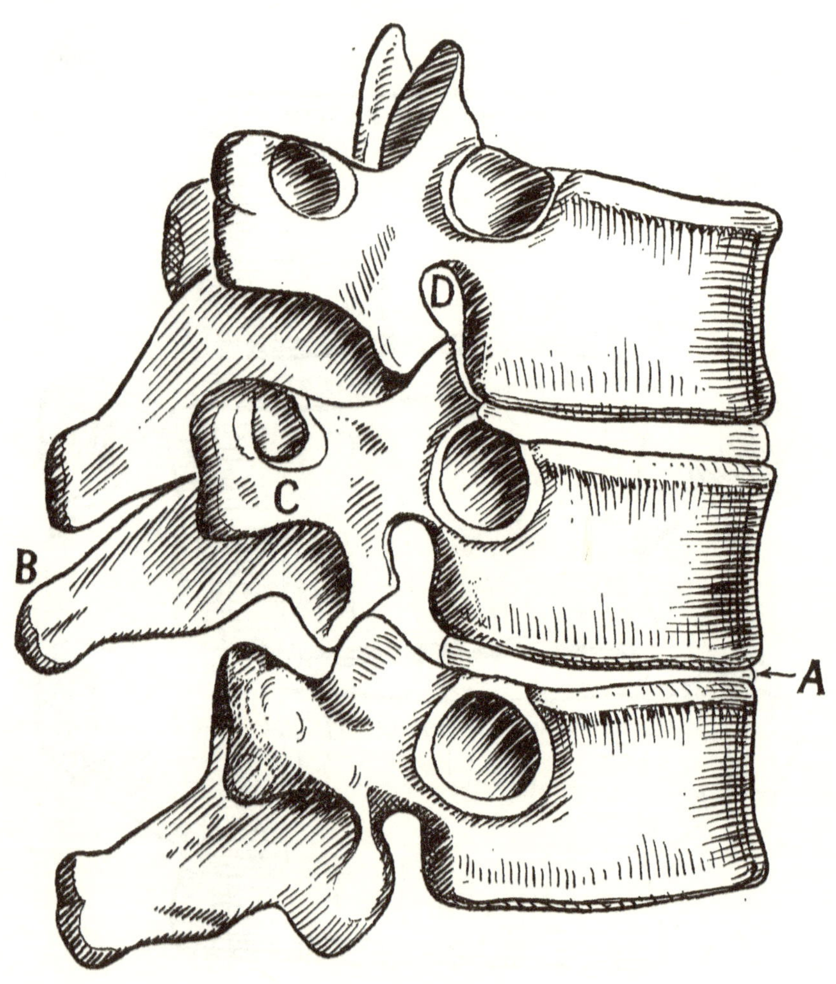

Fig. 4

Antero-inferior subluxation. There is a thinning of the anterior aspect of the disc (A) upon which the subluxated vertebra rests. The spinous process is displaced upward and backward (B) thus bringing it into closer apposition to the spinous process of the vertebra above. The transverse processes are displaced forward and upward (C) and thus are less palpable than those of the adjacent vertebrae and are in closer apposition to the transverse processes of the vertebra above. The superior articular processes are carried forward and upward (D) thus encroaching upon the lumen of the intervertebral foramina.

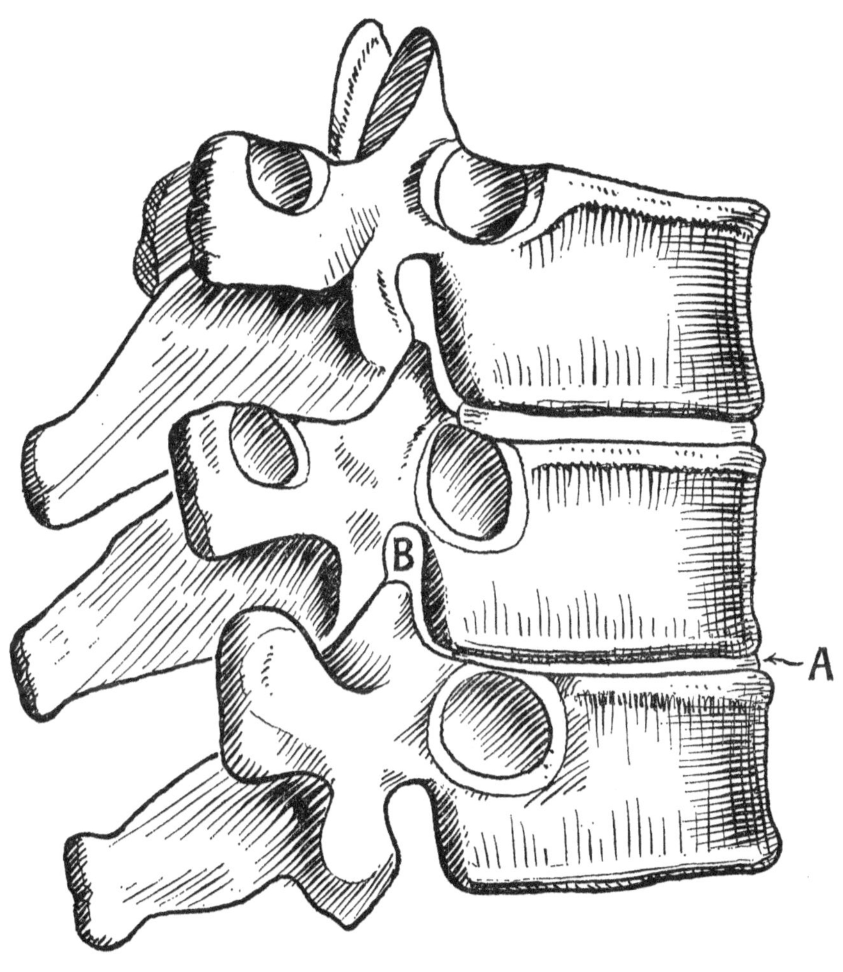

Fig. 5

Inferior subluxation. There is a thinning of the entire intervertebral disc upon which the subluxated vertebra rests (A) causing a decrease in the vertical diameter of the intervertebral
foramen (B) between the two vertebrae.

long analysis of the imperfection of the vertically-held spine from an engineering standpoint, and it is no wonder that mal-alignments occur in view of our habit of using vertically a structure that was designed to function in a horizontal position.

5. WHAT IS A SUBLUXATION?

One should discriminate between the terms subluxation and dislocation. A dislocation is a major lesion of a joint and always involves the tearing of tissue, or major, easily-recognized pathology of skeleton or soft tissue. Major dislocations of spinal vertebrae involve fractures.

On the contrary, a subluxation is simply a slight change in the relative position of adjacent vertebrae. That is, instead of the entire surface area of the bottom or top of a vertebra matching or aligning with the surface of the next vertebra above or below with die-like precision and accuracy, it is moved slightly from that position. There is not an absolute and entire separation between the vertebrae; on the contrary, the greater portions of their upper and lower surfaces still oppose each other—there has simply been a slight shift of one vertebra upon another. This movement can take place in various directions, each producing a different type of subluxation, requiring different chiropractic techniques for correction.

It is because portions of the lower and upper surfaces of the vertebrae form the channels through which the spinal nerves emerge from the spinal cord, on their way to the body organs which these nerves supply and control, that any change in vertebral position (even though slight) changes the shape and size of those channels, and therefore places pressure upon the nerves and their allied structures of blood vessels and lymphatic vessels.

An essential feature of a subluxation is that the vertebra is relatively fixed in its abnormal position and no longer takes part in the normal movements of the spine of which it is a part. As long as a vertebra possesses normal "mobility" (freedom of movement) it is not considered as subluxated. Rigidity in at least part of the spine is found to occur in all ailments or abnormal organic states, but we are so constituted that we often are unaware of that rigidity unless it is of such magnitude as to encompass almost the entire spine. It is, however, evident upon manual examination by individuals trained in the necessary techniques.

It has been found that no section of the spine, however small, can become rigid without having a harmful effect on the portion of the body supplied and controlled by the nerves issuing from that portion of the spine.

6. THE EFFECTS OF SUBLUXATION UPON THE NERVES AND OTHER ASSOCIATED STRUCTURES

(a) Responsiveness and Conductivity of Nerves
When a vertebra is displaced, it encroaches upon the diameter of the channel through which nerves emerge from the spinal column at that point, and causes

pressure upon those nerves. The margins of the vertebrae are not sharp, but rather are smooth and rounded; they do not press directly upon the nerves, but indirectly by pressing upon the tissue surrounding the nerves. This process is termed "impingement". From the foregoing facts it will be seen that the pressure exerted upon the nerves by the displacement of vertebrae does not cut or sever nerves, but repeated experiments have proven that it does impair their *responsiveness* and *conductivity*.

Responsiveness is the property of living tissue to undergo physical and chemical changes when irritated. Irritants, with reference to nerves, include anything which causes nerves to transmit impulses. In fact, nerves function because of their ability to become irritated and to respond to those irritations. Anything that reduces the ability of nerves to become irritated, reduces nerve function. It is a physiological fact that pressure gradually applied to a nerve first increases and later reduces its power to respond to irritants, and finally destroys it. Both increased and decreased irritability result in mal-function and disease in organs affected by the nerves involved. Increased irritability can produce over-activity and spastic conditions; decreased irritability (of nerves) results in under-activity and degenerative changes in the organs affected.

If the power of a nerve to respond to irritants is reduced or lost, it is unable to carry normal impulses, for upon its irritability depends its power to generate impulses and to convey them from or to the brain. Thus, when a vertebral subluxation exists, causing nerve impingement—disease or malfunction follows, because the normal function and condition of every organ of the body is dependent upon adequate nerve functioning.

Sufficient pressure may be applied to a nerve to destroy its irritability and conductivity without injuring the nerve structurally; i.e., pressure insufficient to crush the nerve or to cause any gross, observable damage can yet be sufficient to partially block the impulses which pass along the nerve.

It is positively known:

1. That pressure upon a nerve will impair conduction of impulses by that nerve.

2. That such pressure need be only very slight to impair conduction.

3. That sufficient pressure to produce this effect may be exerted by subluxated vertebrae.

Organs which are deprived of normal nerve impulses, undergo functional or organic changes; their resistance and vitality are lessened, and disease results in the parts affected. The nature of the disease depends upon other contributing factors—the subluxation, by promoting conditions in the organ which make disease possible in that organ, can be considered a primary cause of disease.

The functional activities of all parts of the body depend upon the amount or strength of the nerve impulses received. This has been proven time and time again, by experiments on animals and by other means. If, therefore, anything interferes with the power of conduction of the nerve, the impulses which it normally conveys to the parts which it supplies are then not forthcoming in the required amount or strength, and those parts will suffer correspondingly. There will be either functional derangement, or changes in structure, or both.

(b) Effect of Interference with Blood-supply to Nerves

In addition to the direct effect of pressure upon the nerves, owing to a subluxation, there is also an indirect effect, as a result of partial occlusion of the blood vessels which parallel the nerve paths and which likewise emerge between the spinal segments along with the nerves.

The nerve-fibre requires a constant supply of blood for the maintenance of its irritability. The irritability of the nerve cannot long continue without oxygen; a nerve which has been removed from the body is found to remain irritable longer in oxygen than in air, and longer in air than in an atmosphere which contains no oxygen.

One of the effects of a subluxation is pressure upon arteries and veins that supply the structures of the corresponding segment of the spine with nourishment, and carry away the waste materials. When a constant interference, such as from a subluxation is maintained, then the blood supply is interfered with, causing impoverishment of the nerve cells and accumulation of waste materials, both of which have an adverse effect upon the power of the nerve to function.

Another function of the blood in respect to the nerves is distribution of heat. A nerve which is deprived of this heat loses its power of irritability and conductivity. Obstruction of arteries passing between the spinal segments, resulting from displaced margins of those segments, diminishes the blood supply to the nerves and hence also, the heat which the blood supply normally conveys to them. Impulses transmitted by that nerve accordingly will be impeded, or fail to reach their destination, and disorders in the parts thus deprived of their necessary nerve-force will follow.

Further, the blood has the power to neutralize the acids which are produced by the cells during action, and so maintain the alkalinity essential to the life and activity of the cell. Also, by virtue of the salts which it contains, it secures the osmotic relations which are necessary to the preservation of the normal chemical constitution of the protoplasm of the nerve.

The irritability of nerve protoplasm is markedly influenced by even slight changes in its constitution. If experimentally a nerve be allowed to lie in a liquid of a composition different from its own fluid, and especially if such liquid be injected into its blood vessels, an inter-change of materials takes place which results in alteration of the tissue and a change in its irritability.

If, therefore, the venous flow is obstructed, the acid waste materials of the activities of nerves remain within them, and a change in the constitution of their protoplasm is caused which impairs their irritability and conductivity. The pressure of the displaced margins of a spinal segment produces this effect by obstructing the circulation of the blood in the veins which it transmits.

(c) The Effects of Lymphatics upon Nerves

Another of the physiological effects resulting from a subluxation of a vertebra with consequent narrowing of the space between adjacent spinal segments, is the influence upon the lymphatics, and their effect on the nerves which they serve. The lymphatics which follow the paths of the spinal nerves, have much to do with the metabolism of each segment of the spinal cord. If the nutrition of a certain segment is faulty as a result of an insufficient drainage of lymph, the ability of the nerve to transmit impulses (reflex excitability) is reduced, and the tissues thus deprived of the necessary impulses will fail to function properly.

(d) The Effect of Changes of Tissue Structure induced by Subluxations

Impingements cause irritation which result in a reaction of inflammation. If long continued, the products of the inflammation result in a thickening of tissue or a proliferation of tissue—in either case the added tissue bulk encroaches upon the space required by the structures emerging between the spinal segments, thus increasing the pressure upon nerves, their blood vessels, and their lymphatic vessels.

7. THE EFFECTS OF NERVE IMPINGEMENT, PRODUCED IN ORGANS AND TISSUES TO WHICH THE IMPINGED NERVES LEAD

We have seen that impingement interferes with the transmission of nerve impulses. Any interference with the free and continuous transmission of these impulses; anything which diminishes the power of conductivity of the nerves, is regarded by Chiropractic as an important cause of disease. This science further goes on to state that the factors commonly considered as the causes of disease are not the real cause, but merely secondary factors, and that nerve impingement makes it possible for these secondary factors to be effective in inducing disease. For example, this line of reasoning makes the point that pneumonia toxins are not the primary cause of pneumonia—if they were, then every individual would "catch" pneumonia, since we are all brought into contact with these micro-organisms.

There must, therefore, be something which prevents certain individuals from becoming affected with pneumonia—some factor which, when lacking or deficient, makes it possible for others to contact the disease. This "something" is stated to be the resistance of the former group and the want of such resistance in the latter group.

But what *is* "resistance"? Chiropractic states that resistance is simply another term for normal metabolism, normal functioning of the organs of the body, and a harmonious whole. For the body contains within itself, automatic mechanisms for combating infections, microbes, viruses and other types of invaders, and if these mechanisms are functioning normally, one does not come down with an infectious disease. However, the normal state of the body economy depends upon a free and uninterrupted flow of nerve impulses. Anything, therefore, which interferes with the conductivity of the nerves, or improperly alters the flow of stimuli, must be considered as a primary cause of disease. The place at which this interference usually occurs, has been found by Chiropractic to be at the points where the spinal nerves emerge from the spinal column.

There are many clinical proofs of Chiropractic theory in this regard. For example, it has been found that in nearly all illnesses involving structures of the head and face, such as the eye, ear, nose and throat, subluxations exist in the cervical vertebrae supplying those organs, and that when those vertebrae are adjusted to remove the subluxations, a cure or improvement follows.

A subluxated vertebra may be the *direct* cause of an ailment, in that there are certain ailments which arise directly from the lack of adequate flow of nerve impulses, or it may be an *indirect* cause of illness, in the case of other ailments where a proper supply of nerve energy enables organs or parts of the body to combat or throw off the causes of illness.

The attempt of nature is always to try to preserve health—but nature must have the proper instruments necessary to carry out her designs, if she is to be successful in making repairs. She will be, to some extent, helpless if the nerves that control the affected portion of the body are partially or wholly blocked by impingement.

Chiropractic tests show with much physiological detail the exact manner in which the diminution of nerve flow affects the portions of the body served by the affected nerves. This material is too technical to be included in our present compilation; it can be consulted directly, by any reader who needs proof of the theory involved.

8. THE ROLE OF CHIROPRACTIC ADJUSTMENTS IN THERAPY

Chiropractors learn specific techniques for detecting spinal subluxations and for administering adjustments which have as one of their purposes, that of eliminating subluxations of the spine, and of other joints in the body. The fact that subluxations are universally present in disease, can be proven by spinal X-rays.

Proof of the therapeutic effectiveness of Chiropractic adjustments is of two kinds:

1. *Clinical.* The fact that alleviation or elimination of an ailment occurs in the majority of cases that undergo a series of Chiropractic adjustments.

2. *Instrumental.* Means for measuring the difference between nerve flow before and after spinal adjustments have been administered.

There are two types of instruments which measure the effectiveness of Chiropractic adjustments in improving the flow of nerve energy within the body. One registers heat, the other directly registers nerve energy.

(*a*) *Caloric Indicators*. A spinal subluxation sets up abnormal resistance to nerve flow. This converts some of the nerve energy into heat which is localized at the subluxated area. The first instrument to measure the increased quantities of heat in subluxated areas, was the Neuro-Calometer widely used by the Palmer School of Chiropractors. Similar instruments have more recently been manufactured by a variety of companies.

The Neuro-Calometer and similar instruments consist of a very sensitive thermometer connected to two contact prongs which are applied to the skin adjacent to the spine, with readings taken at each spinal segment. The increase in temperature is quite marked in the same areas which appear subluxated both on X-ray pictures and to the Chiropractor's trained touch. After the adjustment has been administered to relieve the subluxation, a re-check with the heat-detecting instrument shows that the excess of heat has disappeared.

(*b*) *Electrical Indicators*. If a sufficiently sensitive instrument is used with a very high degree of amplification, the flow of nerve energy can be registered directly. It is usually depressed below normal at subluxated areas, and increases after the requisite adjustment of the vertebrae. One such instrument is the Neuro-Micrometer developed by Dr. George A. Wilson.

Chiropractic adjustments have been found to be of benefit in other ways than the elimination of nerve impingement; they also provide stimulation, restore impaired circulation of blood to the spinal nerves in the affected area, and restore the mobility of sections of the spine that have become rigid through prolonged subluxation. Sometimes vertebrae that have been subluxated a long time, continue to resume their subluxated positions after adjustments have been made, yet if mobility of that section of the spine is restored by the adjustments, the patient usually experiences decided improvement in spite of the mal-positioning of the vertebrae. Rigidity and lack of elasticity are characteristic of aging and decay. Any measure that can counteract them, works in the direction of better health.

There seems to be a widespread popular misconception that chiropractors receive short courses of instructions and therefore are not adequately trained to handle serious ailments. Nothing could be farther from the truth, regarding present-day conditions. The course for Medical physicians takes four years, of which one year is mainly devoted to drugs. The course in most of the forty-eight states required for Chiropractors before they can receive a licence to practice, comprises four intensive years of instruction equivalent to the Medical course in the basic sciences, anatomy, pathology, physiology, etc.; in fact it is equivalent to a Medical course in every field except that of drugs, which Chiropractors do not use and do not believe in. And, in the field of Neurology, the training received by Chiropractors is far superior to that of any medical physicians except possibly the Neurology specialists.

It was only at the very start of Chiropractic that short courses prevailed—the same situation applied to medicine in *its* early days.

9. CAUSES OF SUBLUXATIONS

In view of the very important role which subluxations have been shown to play in various types of illness, a review of the conditions that cause subluxations may be of interest.

These causes fall into two categories—external and internal.

External causes include jars, falls, twists, jolts, etc. of a force strong enough to overcome the elasticity of the ligaments surrounding the vertebrae and of the discs between the vertebrae. These may be due to occupation, habits or injuries. Very often, a persons does not realize at the time, that an injury has been sufficiently severe to cause a spinal subluxation. A complaint arising as a result of a subluxation may not evidence itself to a patient until a considerable period of time has elapsed, and very often the patient's mind does not connect up the injury with the complaint. In some individuals, jars, falls, blows, strains and twists may be so slight as not to be painful or to cause any recollection, and yet be sufficient to cause spinal subluxations.

In old age, the vertebral column gradually becomes shortened, and less straight. This degenerative process also is likely to produce subluxations.

Internal causes arise through what is known as the "reflex arc" or more briefly, just "reflexes". Any ailment arising in any organ or body tissue, results in an irritation which is conveyed along the nerves to the spine. These irritations tend to induce spasms in the muscles and ligaments that hold the spinal segments in place, and these spasms frequently result in pulling the respective spinal segments out of their proper positions, causing subluxations. Therefore we see that not only can subluxations cause disease through reducing the nerve supply to the affected area, but they also can start as a result of disease and then become perpetuating factors which prolong the disease, or transform it from an acute to a chronic nature.

COMMENT ON CHIROPRACTIC RESULTS

It should first be noted that within the field of Chiropractic there are several schools of thought. One school typified by the Palmer graduates, is that the top cervical vertebra named the "Atlas", which handles the nerve supply to the brain, is the most important vertebra to adjust, inasmuch as the brain plays such a large part in controlling the rest of the body. The extremist faction in this group, known as the H-I-O (Hole in One) people, claim that it is *never* necessary to adjust any vertebra but the Atlas. These viewpoints are vigorously opposed by the rest of the Chiropractors, who find that better results are obtained on a larger percentage of cases by adjusting all vertebrae that are subluxated. So we have the division between the Atlas Chiropractors on the one hand, and the general chiropractors on the other hand. The group of general Chiropractors is further split into two factions—the "straights" and the "mixers". The "straights" believe that the only desirable therapeutic method is the administration of

Chiropractic adjustments. They do not advocate or use any other method of therapy. The "mixers" use Chiropractic adjustments, in combination with or in addition to various methods of physical therapy which seem indicated in individual cases, and also advocates certain forms of diet, and sometimes the use of food supplements.

The auxiliary methods used by the "mixers" can be grouped into two classes:

(A) Auxiliary methods having an effect upon the patient which aids the Chiropractor in administering satisfactory spinal adjustments.

These include: massage, infra-red rays, short-wave diathermy, and various reflex methods—the primary purpose of these methods is to secure a greater degree of relaxation so that the spinal adjustments can be more easily made and will be more effective.

(B) Auxiliary methods to aid in re-establishment of the patient's health, supplementary to the adjustments.

These include: sine-wave and galvanic currents, ultra-violet irradiations, soft-tissue manipulations, radionics, colonic irrigations, diet, vitamin therapy, and other nutritional supplements, etc.

There seems to be no doubt that Chiropractors have cured or brought great improvement to many people who had not obtained satisfactory results from medical and other forms of treatment. Percentage of satisfactory results with Chiropractic varies from one office to another. This is partly due to differences in individual ability of the Chiropractor—a factor involved in the effectiveness of any form of therapy; and partly because of the large variety of techniques of adjustment, coupled with the fact that some Chiropractors use a larger number of techniques than others, and may be more proficient in selecting the best techniques applicable to specific cases that come to them. For example, one of the standard chiropractic text books lists and describes seventy-one different techniques for adjusting various portions of the spine, twenty-four techniques for adjusting the shoulder, seventeen techniques for adjusting the foot, and forty-four techniques applicable to other joints in the body; a total of 156 different techniques of adjustment. No one man can be equally skilful in the use of all these techniques, and it is inevitable that some practitioners will stress certain of these techniques while other practitioners will prefer other techniques. This perhaps explains why a patient may find that Chiropractor X could not help, but Chiropractor Y's treatments were successful, whereas patient B may have just the reverse experience and find that Chiro X took care of a condition which did not yield to Chiro Y. The practice of Chiropractic is a highly individual matter, and depends for its success partly in matching the right techniques to the patient, and applying them with the particular degree of force and speed most suited to that particular patient. The effect of the same adjustment technique used by different Chiropractors can vary in accordance with three factors:

1. The degree of skill in obtaining just the right contact and applying the adjusting "thrust" at exactly the correct angle.

2. The degree of force applied. The amount of force needed to achieve a particular adjustment varies with different patients.

3. The degree of speed used in the adjusting technique. A high-speed adjustment has different effects upon the nerves than does a lower-speed adjustment.

Failure of Chiropractic to achieve satisfactory results in a minority of cases, is frequently due to a pressing need of the patient for a more balanced nutrition and body chemistry, or may be due to the need for psychological treatment. Realizing this, some Chiropractors are adding these types of therapy to their practice.

BIO-DYNAMICS – THE NEW CHIROPRACTIC

WHAT IS DEATH?

NEURO-MICROMETER measurements show that it is the brain cells which deteriorate first—terminating the control and co-ordination of organ functioning. When this co-ordination ceases, the body undergoes the change which we call "death", even though the Neuro-Micrometer measurements show that the body cells (in contrast to the brain cells) remain alive for many days and even weeks after the so-called "death". In fact, it is a common occurrence for body cells to register higher states of activity for a time after death than before death.

It is found that the cells continue to defend their life integrity until they die either of starvation or of poisoning from their own excreta. These processes occur sooner, in those persons whose cells have been weakened by disease. When a healthy person is killed in an accident, the measurements show that the cells can live for months before ceasing their activities. This explains why hair and nails can continue to grow after death.

In those who die from disease, Neuro-Micrometer readings show an abnormal state of alkalinity just before death, except where death occurs from heart attack, loss of blood or asphyxiation. The alkalinity neutralizes the polarity between the brain and the body, thus disrupting the mechanism by which organ functions are co-ordinated.

In life, the flow of impulses from the brain to the body (efferent impulses) involves a greater quantity of energy than the afferent impulses from the body to the brain. That means the brain cells must work at a higher rate of activity than any other cells in the body, in order to maintain the functions of control and co-ordination.

The brain deteriorates first, because it is composed of more highly organized, more sensitive tissue; is more susceptible to adverse influences, and is forced to function at a higher rate of activity than any other part of the body.

WHAT IS LIFE?

Dr. Wilson presents the concept, derived from electronic observations of the human body with the Neuro-Micrometer, that:

> Life activity is a series of responses to a series of demands, and exists only so long as the body is able to respond to the demands.

The types of responses the body makes, fall into three categories: Instincts Urges, and Cell-Group Responses.

Instinctive Responses. These come into action in defence of the total organism against an adverse environment, such as enemies or bad weather. This type of response involves the total organism, and ceases at death.

Urges, to which the body responds, are the needs for nutrition, protection and reproduction. Death likewise terminates the response to these urges.

The two groups of responses listed above, are concerned with the external protection and perpetuation of the body. On the other hand,

Cell-group responses are concerned with maintaining health through proper co-ordination of the body's internal functional activities, including:

> The elimination of internal toxins.
> Maintaining the acid-alkaline balance.
> Repelling the invasion of harmful micro-organisms (bacteria, viruses, parasites, etc.).
> Combating the effects of injury, whether from internal or external causes.

It is a fascinating finding of the Wilson research, that this group of responses continues after the death of the organism as a whole—the cells of the body do not die until later.

Instincts and Urges are *general* patterns of response. In contrast, Cell-Group Responses are of a *specific* nature and derive from inherited defence patterns, inherent in all life.

The specific nature of this type of response can be illustrated by such facts as the following:

> The stomach and digestive systems go into action only when food has been eaten.
> The heart beats faster only when a person runs, climbs stairs, or becomes excited.
> Likewise the liver, kidneys, bowels, pancreas and other glands go into action when there is work for them to do.

Thus when there is no demand, there is no response; and when there is no need for a response, the body reverts to and remains in a state of functional rest, except for residual (metabolic) activities.

In going into action after exposure to adverse factors such as toxins, excess acids or alkalies, harmful micro-organisms, or the effect of injuries, a tissue (cell-group) is defending its life integrity.

It should be obvious, therefore, that it is the response of cell-groups that protects the body from the adverse effects of injuries or attacks of disease. Disease gains a foothold only if the cell groups are not strong enough to respond adequately to such demands. Inadequate response can occur because of physical weakness—which essentially means cell deterioration—or it can occur when the demands are too great to be met.

Of the three types of responses which the body can make, it is the Cell-Group

Responses which are particularly important in conditions of disease and injury. Therefore it is of great importance to explore this type of response in detail so that Doctors will have a better understanding of means to evoke, accentuate and maintain this type of response when needed for recovery from illness. This exploration will be the subject of following chapters.

The further one goes into the subject, the more one realizes that man is more than mere "energized matter". The energy which activates the body, does not act blindly; intelligent direction is evident here, as in all the cosmos and in all manifestations of life.

WHAT IS DISEASE?

When a person becomes sick, we have looked upon his ailment as being the thing which made him sick. Instead, we now say it is the other way around—sickness is only the effect of the body's efforts to rid itself of factors detrimental to life and health.

From this point of view, the nature of disease can be summed up in a few words: disease does not come from what "something" does to the body as has so generally been accepted, but rather from what the body does to that "something".

It has been recognized in the past that, when a person takes a physic, it is not the physic which moves the bowels but the bowels which moves the physic. Yet the full significance of what took place had not been understood. In the light of our Neuro-Micrometer research, it can now be recognized that the physic simply creates a demand, to which the body responds by moving the bowels and all else in them.

Similar actions can take place in other parts of the body, from other causes. This line of investigation, carried to its logical conclusion, produces the answer that disease in reality is nothing more than *the way in which the body responds to certain demands!* The name of the disease comes from the manner in which the response operates; the manner in which the response operates is determined by the weakness or susceptibility of the parts involved; and the weakness or susceptibility or the parts involved are frequently the result of interference to the flow of nerve energy to those parts.

Thus, what is called sickness is actually intended to be a beneficial process, such as a boil or abscess to get rid of pus, a fever to burn up body poisons, diarrhoea to eliminate other poisons, an enlarged heart or liver which became enlarged to help with the extra work it is called upon to perform, etc. The processes which we term disease, *become destructive only when the response is allowed to continue unabated or the demand becomes too prolonged.*

In other words, a sickness should run a self-limited course and come to an end. Wrong treatment, or the lack of appropriate treatment, can result in the sickness becoming an ailment and in the ailment becoming a disease. Incorrect treatment, frequently administered by those who do not understand the

DEMAND-RESPONSE concept, can cause the demand to become prolonged or the response to continue unabated for an indefinite period. In most cases the proper correction of nerve interference can prevent both of these adverse consequences.

Factors which can create prolonged demands, are injuries, adverse weather, and body poisons such as toxins, excess acids or excess alkalines. It is the factor of body poisons, which is most frequently encountered. Neuro-Micrometer research shows that these are universally present in all ailments.

Thus, body poisons, more than any other factor, cause the demands made upon the body to become prolonged. Such poisons can come from over-eating, overwork, the eating of wrong foods or wrong food combinations, insufficient rest, nervous tension, emotional upsets, medication and wrong treatments, as well as from sluggish elimination. It is factors such as these, that force the body to continue a response to the point and degree that becomes abnormal. And when that point is reached, what started out as a constructive process, now becomes destructive. If such a person dies, it would be the body in reality which destroyed itself, because excessive demands forced it to respond so excessively that its response got out of control and became destructive.

There are two sides to the Demand-Response conception of the nature of disease. On the one hand there are the demands which become excessive as we have already discussed—on the other hand the response originally elicited to meet a demand, may continue unabated after the demand has been satisfied. A response continued beyond its need, likewise produces abnormal or diseased conditions in the body. Unabated response comes mainly from nerve pressure and twenty-nine other factors which can cause interference to the flow of nerve energy, and therefore prevents the proper co-ordination of body functions from being maintained.

Traditional chiropractic has placed great emphasis on the role of nerve pressure in interfering with nerve impulses from the brain to the body organs and tissues—impulses to step up their activities in response to demands resulting from sickness. Dr. Wilson's research shows the need for more attention to removal of interference with the other type of nerve impulses—*from* the body organs and tissue *to* the brain. It is this type of nerve impulse which must give the "cease-fire" order, to direct the brain to cancel the response of excess activity in a particular organ or group of organs or body tissues, when the demand has been met. Nerve interference can prevent the "cease-fire" order from reaching the brain, hence the over-activity of parts of the body continues unabated long after the need for such over-activity has vanished, and the result of this set of consequences is that either the functions or the cell structure "go wild".

If it is the functions that run riot, we have ailments such as choria, palsy, dysentery, tachycardia (rapid heart beat), etc. If it is the cells which go wild, then tumours, anaemia, leukaemia, tuberculosis and cancer and many other ailments can develop. Why? Because the cells are then no longer under the control of the cerebro-spinal nervous system and a directive intelligence.

Thus the body, when its cells or functions go wild, can and too often does destroy itself. The process of destruction which takes place is called a certain sickness, ailment or disease, and is named according to the direction it takes and the tissues involved. Nevertheless it represents the body's efforts to get rid of an unhealthy condition, and in which its efforts go out of control and go wild.

Sickness comes not by chance, but for a purpose, and keeps returning, often under different names, until that purpose is fulfilled. Its purpose is to overcome conditions detrimental to life and health.

To recapitulate: Sickness or disease is an aroused response to meet a stepped-up demand and only becomes destructive when the demand becomes prolonged or the response is allowed to continue unabated.

WHAT IS HEALTH?

The popular answer to this question is usually: "Health means to feel good."

A scientific explanation is not so simple nor so easily obtained. It must include an explanation of body structure, of body functions, and of a directive power—all of these are factors in health.

A person can have health only if:

(1) the body structure has reached a high degree of development, and

(2) the body functions have attained a high degree of co-ordination, intelligently controlled, through which life in all its meaning and purposes can be expressed.

1. *Body Structure*

Much is known in this field, concerned structure, body chemistry, and microscopic as well as gross aspects of the muscles, ligaments, nerves and bones, the organs, glands and tissues, the different types of blood cells and the cells of all parts of the body. These have all been named, described, located, and their purposes delineated.

These facts are common knowledge among all doctors, including those of our profession. From there on we, as chiropractors, have thought that life of the cells came from nerve energy. Philosophically that idea is tenable—scientifically, no.

The work of Carrel with the chicken's heart; of the Ralston Club with pieces of bone, skin, muscles and chicken hearts; and of those who carry on tissue culture for surgical purposes, in which different parts of the body were and are today kept alive for indefinite periods of time—all show how scientifically untenable is the concept that nerve energy is the life of the cells. For, in none of these experiments was nerve energy involved, yet the cells lived.

Cells are not kept alive by nerve energy—rather, the cells generate their own life; the function of nerve energy is *co-ordination*, as will be outlined shortly.

At this point we are concerned with the health of the cell—the primary life unit of the organism.

So important are the cells in the life and activity of the body that we could say; as the cells, so is health. The fact of the matter is that no matter what the ailment, it has but little adverse effect on the body until poisons from it affect the cells.

The level of cell-life and health, as Neuro-Micrometer research has revealed, bear a close relationship. The lower the cell-life, the more precarious is the health. It is found that patients do not live longer than ten days after their cell-life reading declines below 34. Of those whose cell-life registers below 78, none have survived longer than three months. Of those whose cell life has declined below ninety but is above eighty, it is possible to restore many to good health. Thus it is obvious that the level of cell health is of signal importance to the health of the organism as a whole. With the Neuro-Micrometer, the level of health of the body cells can be accurately measured, and close track kept of the effect of therapeutic measures. This opens wide new vistas for the improvement of the healing arts.

2. *Functional Co-ordination*

Nerve energy, through the cerebrospinal system, controls and co-ordinates the activities of the digestive and eliminative systems, to the end that the cells are amply supplied with the necessary nutriment and their excreta taken care of. In fact, all of the body's functional activities are controlled by nerve energy created in the cells of the brain. Thus the brain, besides carrying on the function of thinking, must also create the energy needed for the control and co-ordination of all bodily functions, so it is not surprising that the cyclic action of the brain cells is around 600 per min., while that of cells at the body's periphery approximates 72 per min.

It is this higher rate of activity which wears out the brain cells first, and why death is the cessation of the co-ordination of functions, as was mentioned in the first article of this series.

In experiments at Yale University, the heart, lungs, stomach, bowels and kidneys were taken out of a cat, and kept alive in warm water by supplying artificial respiration and artificial circulation of blood. Such cats were reduced to a mass of protoplasm and could no more carry on an independent existence than could the chicken heart which Carrel kept alive for twenty-eight years.

All of which but emphasizes the body's need for functional co-ordination or it would regress to a mass of protoplasm and become totally dependent. The only source of co-ordination is through the cerebrospinal nervous system, and such co-ordination can be exercised properly only when the nervous system is free of interference. This illuminates the real role of chiropractic.

On the basis of the new facts at hand, obtained through electronic research, Dr. Wilson's definition of health follows:

"Health is a finesse of balance between demand and response, controlled and co-ordinated by an intelligence through the cerebro-spinal system of nerves, in which the body's structure, functions, chemistry and neuro-electric energy are maintained in amounts and degrees neither too much nor too little but sufficiently to give the greatest latitude to the expression of life, thought and action."

CELL RESPONSES AND THE BODY'S LINES OF DEFENCE

Studies in the past concerning anatomy, functions and energies of the living person, have been made mainly on a gross basis (in the large—regarding whole organs or parts of organs) rather than on a microscopic or cell-level basis. Histology, it is true, goes into a micro-study of the cells, but it has been oriented to *physiology rather than to bio-dynamics*. The latter gives a better insight into cell life and cell functions.

Thus in order to understand the Demand-Response factors, it is necessary to have a better understanding of the genesis of life—the cell; both from the standpoint of cell-structure which is studied in cell biology, and from the standpoint of cell-functions, which are studied in cell biodynamics.

From a study of cell biology we learn, as discussed in a previous article, that the cell can live independently of nerve energy, so long as the environment it requires is adequately maintained. Such an environment requires nutrition, waste disposal, oxygen and warmth. A study of cell biodynamics reveals how the cells are able to create their own "spark" of life and to adapt themselves to changing conditions, the things necessary for their perpetuation.

Inasmuch as space will not permit more than a brief discussion of the factors micro-anatomy has revealed concerning the cell, we shall only mention the high spots. They are:

(1) Cells can live in a healthy state in a normal environment indefinitely.

(2) Nothing can be added, beneficially, to the normal cell environment.

(3) Cell groups (i.e., groups of organized cells) inherit a pattern of defence.

(4) The only thing that can cause a cell group to defend itself is an adverse environment.

(5) *Any deviation in the structure or function of cells is but evidence of their efforts to adapt themselves to an abnormal environment.*

(6) The factors which cause cell environment to change adversely are heat, cold, body poisons and injury.

(7) Heat, cold and body poisons put the cells in a state of distress; which they must change through cell-group action, or adapt themselves to it or die.

(8) Injury puts the cells in a state of stress; again they must adapt themselves, meet the issue or die.

(9) Thus when cells change their nature, size or behaviour, it is because of stress or distress and these come from adverse changes in cell environment. Cells do not think. They have no thinking mechanism. They can only act on the basis of an inherited pattern of response, and the pattern of response is to defend their life integrity against destructive factors. Thus it is only when something threatens their life that cells or cell-groups go into action. Such is the defensive side of cell-groups. Their offensive side involves cell bio-dynamics; to be discussed in the next article.

Cell Adaptation

Nothing equals the adaptive ability of cells, which but indicates their ability to meet demands. A person can go where the heat may reach 130° or go where the temperature may be 50° below zero, yet in each case his body's temperature will remain at its usual level—98.6; because the body is able to adapt itself to such temperature variation. Such is the body's power to meet demands, and the following are other ways in which the body responds to other types of demands. If you eat something injurious, and your body is functioning normally, it will either vomit the offending stuff up, cause a diarrhoea or a polyuria, according to the nature of the ingested substance, in order to get rid of it. If you run, your heart and lungs will work faster in order to supply the oxygen the body needs and to circulate it faster to all parts of the body.

Such types of response are made possible by the concerted action of the cells, organized into cell-groups. Can anything be more marvellous than the body's ability to meet such a variety of demands? Each time a demand presents itself, the body will produce the correct response necessary to take care of it—unless it is out of the range of the body's ability to meet it, or if there is nerve interference.

Thus the body is ready and capable of meeting and successfully responding to any and all demands within its range. If an injury or sickness happens to take place because of something over which the body has little or no control, it should run a self-limited course. If it doesn't, it will be because of nerve interference or because of a bungling by someone. Thus a chronic ailment is a bungled ailment or is an ailment in which the nerve interference has not been lifted. No ailment becomes chronic, ever (states Dr. Wilson) unless one or both of these factors are present. Otherwise the body can and would regulate its own affairs.

The Body's Four Lines of Defence

The body's four lines of defence are important factors in enabling the body to meet any and all demands encountered within its range.

Its lines of defence are:

1. The organs of elimination.
2. Fevers.

3. The lymphatic system.

4. The necroses.

1. *Elimination*

The organs of elimination are, of course, the bowels, kidneys, skin and lungs. When they work normally, and a person does not overeat, overwork, eat wrong foods or bad food combinations, or have emotional upsets, the elimination organs will keep the body free of toxins, excess acids or alkalines—the body poisons. But when the job is too big for the organs of elimination to handle successfully, or if they are unable to successfully handle the situation because of nerve pressure, body poisons will accumulate and become a contributing factor in the development of sickness. In our research at Spears, we have failed to find a single sick person who hasn't a high degree of body poisons. Thus such poisons are a significant factor in all sickness. In fact, without their presence it appears that there would be no sick people. More important—we have found that the level of body poisons and the level of pathological activity parallel each other. When body poisons go up, so does pathological activity; and we can't materially decrease such activity until we get rid of the body poisons.

2. *Fevers*

When the eliminative organs are unable to successfully keep the body free of such poisons, the body responds to the demand by creating a fever through which to burn them up. The mechanism of a fever is an interesting study. It has been considered to be an oxidative process through which the poisons are "burnt" up. But we have found other factors to be involved. To illustrate; if the body's reserve energy, as we call it, registers 95 points on the Neuro-Micrometer, it represents 4,750,000 cytomicrovolt units of cell life. Then if the body's functional activity as we call it, registers 90 points, it will represent 4,500,000 neuro-microampere units of cell-group activity, meaning the body is functioning at that rate of activity. In such an event, the body's amperage will be lower than its voltage and in such a situation only a normal heat is produced. However with the unit of cell life still at 4,750,000 or at not more than 5,000,000 cytomicrovolts, its maximum point, and the neuro-microamperes go up to 10,000,000 units, its highest point—the body's amperage will be twice as great as its voltage, and that creates heat.

This oddly enough, has been found to be true as a result of the use of the electric chair for the execution of criminals, and also as a result of the development of electric welding. The first criminal to be executed, when a higher amperage and a lower voltage was used, was roasted to death. Now, as a result of experiments, the voltage for executions is set at 2,000 and the amperage at 11. The high voltage and low amperage destroys the nervous system but does not roast its victims. On the other hand, with one volt and from 5,000 to 50,000 amperes, the intense heat necessary for electric welding is developed. Thus a fever but represents the body's response to a demand in which its amperage—

which comes from the degree to which the body is forced to function—exceeds its voltage and causes excess heat. When the poisons are "burnt" up, the body's functional activity returns to normal. The amperage drops and the fever stops. The fever is then said to have run its course.

3. *The Lymphatic System*

Under normal conditions the lymphatic system maintains a certain degree of activity, at a rate no higher than the level of metabolic action necessary to maintain the body's basic functions. Neuro-Micrometer measurements of the micro-electric variations in the potentials of the lymphatic glands, show that the lymph system is forced into a high rate of activity by the presence of disease toxins. These toxins are a product of morbid pathology. The presence of such morbid toxins are also reflected in the blood sedimentation rate, and in the "blood-cell clumping pattern"—a new diagnostic technique which will be described later.

A test for the presence of morbid toxins is becoming a simple method of screening or classifying patients as to the seriousness of their ailments and the type of ailment present. This is an important factor in diagnosis.

4. *The Necroses*

The necroses represent the conditions through which the body establishes external drainages in its efforts to get rid of morbid toxins. In a minor sense, pimples, boils and abscesses come under this heading. However, we refer more to running sores. They but represent the body's response to a demand, in which it uses the fourth line of defence to get rid of factors detrimental to life and health. Thus, the fourth line of defence, as is true of the other lines of defence, is also an important diagnostic factor.

PRINCIPLES REVEALED BY THE DEMAND-RESPONSE FACTORS

A study of the Demand-Response factors reveals not only that the body does nothing except because of a demand and in relationship to a demand, unless functions go wild; that it is not what "something" does to the body, but what the body does to that "something", that causes sickness, disease or death; and that mentally and physically we are but cross-sections of the demands made on our minds and bodies and their ability to meet such demands; but it also reveals that it is not what we do to a patient that gets him well, although what we do is very important; it is what the body does about what we do. When the body's ability or power to respond to what we do is high, the patient gets well; otherwise what we do has little or no effect.

CO-ORDINATION OF CELL FUNCTIONS

A few years ago, I (quoting Dr. Wilson) addressed a group of medical students and interns. In my talk I stressed four fundamental factors: (1) that the cells are autonomic units and can live independently of the body or of any supply of nerve energy so long as they are supplied with food, their waste material taken care of, and they are kept in a suitable environment; (2) that when they become organized into cell-groups and the cell-groups into a more complex organism, as in the higher forms of life, the cells lose much of their autonomic independence; (3) that when such as this takes place, the group activities of the cells must be controlled in relationship to themselves and co-ordinated in relationship to other cell-groups; otherwise higher forms of life never would have been possible; and (4) that it is through the cerebrospinal system of nerves that such controls and co-ordinations are maintained.

At the close of my talk, a medical instructor asked this question: "I presume doctor, that you are acquainted with the work of Speransky, and that he performed operations on some cats and dogs in which their brains were removed; and also, that these animals continued to live a normal life even to having offspring?" Replying that I was quite familiar with the work of Speransky, I requested permission to ask a question of him before answering his question. He was willing. "Doctor", I asked, "did Speransky cut out the medulla, the pons, the caudate nucleus, the corpus collosum or the thalamus?" He had no more to say. Those are the parts of what has been termed the old brain, and it is the old brain, located at the base of the skull, that controls all involuntary functions. Cut the old brain out and there could be no co-ordinated life. Such an animal would then be reduced to nothing more than a mass of protoplasm and could not long survive. Such as this happened when Cannon of Yale University experimented with cats. Instead of cutting out the brain, he cut all the nerve connections between the visceral organs and the cerebrospinal nerves; as we discussed in a previous article of this series. The thing which happened to those cats is, to a lesser degree, what would happen if the old brain had been removed. The cats were made totally incapable of caring for themselves and could continue to live only as someone or something would take care of their needs. What a life!

FACTORS IN EVOLUTION

When the factors which became organized into the cells, acquired or were given the ability to respond to internal demands, microscopic units of life—the cells—came into being. But it was only when the cells acquired or were given the ability to meet external demands by organizing into cell-groups and the cell-groups into more complex organisms, that higher forms of life could come into existence. Higher forms of life posed new problems which had to be met. Such problems involved the question of cell-groups being able to work harmoniously together, each working for the good of all; and not unlike the development of civilization—as it should be.

Thus a point that has been overlooked in the development of higher levels of life is that, concurrently with the evolution of life, there had to be also evolved a system of co-ordination—a system of law and order in the body—else there could have been no higher forms of life. For a complex organism could no more have been possible without adequate co-ordination, than there could be a civilization without law and order. The system of co-ordination in the body is, of course, the nervous system. There is something about the nervous system that the medical profession has overlooked and, I fear, too many of our profession are doing the same thing. It is this: that health is a complete co-ordination of all body activities into a smooth-running organism; that sickness is a disruption of co-ordinated functions, in which the body no longer can control its own activities; and that death is a complete cessation of all function co-ordination.

When there is no Co-ordination

Our medical friends do not recognize a need for a co-ordinative power in the body, or if they do, they believe it is maintained by the glands and that the glands are controlled by medicines and hormones. Again they are wrong, as will be discussed in the next article in which we will discuss nerve energy and how it flows, and its part in the Demand-Response factors.

Let us go back for a moment to the Cannon experiment. He proved that cats can live without a co-ordinative power; yet—so have experiments in tissue culture (the keeping of tissues alive for use in surgery) proved the same thing, as did Carrel's experiment with the chicken's heart. To all this we can say—so what! and we would be right. But we need to know the real answer. We need to know the status of these types of life; to know that they are thus reduced to the basic status of life—the level of cell life; and that as such, they can no longer forage for the necessities of life nor defend themselves against enemies or destructive forces. The fact of the matter is, and as was mentioned a moment ago, that life evolved only as a system of cell-group co-ordination was developed, and life is maintained normally only as the co-ordination is adequate. Conversely stated: to the extent that the co-ordinative power is curtailed, the functional activities of the body will get out from under control. This means sickness or disease. If there is further curtailment, the co-ordination needed through which to maintain the life of cell-groups would cease and death would ensue. Thus life goes up to higher levels or reaches a higher status only as each of the parts of the medium through which it is expressed is co-ordinated to a nicety, and regresses to a lower level as the co-ordination becomes disrupted. If the disruption of function co-ordination which comes from an interference to nerve flow, continues on its downward course, all that normal co-ordination has given to the body—its shape, its size, its freedom of operation and its healthy state—becomes lost, and the body eventually returns to an unco-ordinated mass of protoplasm as did the cats, and as do the pieces of flesh, bone and nerves kept alive by tissue culture. Thus the difference between the lowest form of life—a mass of protoplasm, and the highest form of life—man, is the co-ordination of functions brought about by the ability of the cells as individuals and as cell-groups, to respond to both internal and external demands.

RESPONSE TO A DEMAND

When my research revealed that the body or any of its parts functions only when there is work to be done, otherwise it reverts to and remains in a state of functional rest; and that it changes from such a state only because of and in relationship to a demand, it laid the foundation of an understanding of the Demand-Response factors.

In this article we have stressed more the part that Demand-Response factors played in the evolution of life and the part it plays in health. In a previous article—What Is Disease?—we discussed a new concept of disease, viz., that it is caused by the body's response to a demand in which the response gets out from under control, and that disease gets its name from the particular tissue involved and the degree of the involvement, instead of from its being caused by a different factor or set of factors. We have also said that death comes only when the body can no longer respond to any demand, except when it comes as a result of an accident. In the following article we will discuss other important factors.

NERVE ENERGY, CELL ELECTRICITY, AND BODY POLARITIES

When Carrel kept a chicken's heart alive for twenty-eight years, he made history. His experiment forced a change in our concept of the nature of life, the nature of the body's functional activities, the nature of nerve energy and the part it plays in body functions—just as Einstein's theory of relativity and Plunck's quantum theory forced a change in the concepts of scientists concerning the nature of matter and of energy. Whereas, before Carrel's experiment we had thought nerve energy to be the life of the body, we now found the life of the body was the blood. Whereas we had thought there was as much nerve energy in a sick body as in a well body, we now knew there was only as much energy as the body was capable of creating. And whereas we had thought no functional activity could take place in the body without nerve energy from the cerebrospinal system to carry it out, we now knew nerve energy to be the factor which co-ordinates all the body's functional activities into a smooth-running organism and that the body or any of its organs, glands or tissues function except because of and in relationship to a demand. Thus when these factors became known, the stage was set for the development of the Demand-Response factors, from which we have gained a more scientific understanding of how and why the body func- tions.

Before we discuss further aspects of the body's Demand-Response factors, however, we should have a brief summary of the factors that Carrel's experiment with the chicken's heart revealed.

CARREL'S EXPERIMENT

Herewith is my analysis of the factors which Carrel's experiment revealed:

(1) That cells are individual, autonomic units and can live independently of the body or of any supply of nerve energy.

(2) That cells are capable of maintaining their own individual existence so long as their requirements for nutrition, elimination, oxygen, warmth, protection and reproduction are adequately met.

(3) That cells have no inherent intelligence within the meaning of the term, even though in some schools of thought an individual intelligence has been ascribed to each cell.

(4) That while cells can and do take care of their individual demands in response to their own needs, they are incapable of responding as individuals to demands arising away from themselves which involve the activities of cell groups. This is a function of the brain, which is a group of organized cells.

(5) That the brain, besides having the functions of mentation and of co-ordinating body functions, also has the function of creating—through its cells—the energy necessary through which to co-ordinate all body activities.

(6) That the cells lose much of their autonomic independence when they become organized into cell groups, and make group responses only as and when directed and controlled through the cerebrospinal system of nerves by the co-ordinative power of a directive intelligence.

(7) That when cells are not adequately controlled, their growth of group functions can and frequently does "go wild".

(8) That the ability of cells to live independently of nerve energy emphasizes the fact that they are micro-dynamos and can supply their own individual spark of life. Thus they are their own source of energy. This does not preclude the probability of there being a "breath of life" spirit or soul. We are dealing only with the physical factors of life.

(9) That when the cells are their own source of energy and it is the nutrient in the blood that keeps them alive, the blood is the life of the body.

WHAT IS NERVE ENERGY?

Since it is energy that makes possible the body's power of response (its power to respond to a demand), and it is over the nerves that this energy is conveyed to all parts of the body—we should know more about nerve energy, what it is, how it is generated, and how it flows over the nerves.

Nerve energy has been found to be bioelectricity. There is no difference in "essence" between it and commercial electricity. The difference lies only in the medium through which it is conveyed—the nerves—and in its method of creation or generation.

Nuclear physics, the science that has developed atomic energy and produced the atomic bomb, supplies the answers to our questions about electricity. By the term "nuclear" we do not refer to the nuclei of cells of the body but to the nucleus of the atom—the atoms of which the body cells are composed.

"Electricity, as we learn from nuclear physics", to quote from my article in the August *Spears Sanigram*, page 2, "has no more permanent manifestation than has steam, fire or wind, which are only actions and reactions. These owe their existence to a combination of factors and cease to exist when such combinations cease. The same thing is true of electricity. Instead of being a physical something, electricity is only electrons in accelerated action. When not in such action the electrons revert to their former state in which they remain until their action is again accelerated.

"Unlike electromagnetic induction, which is more often used in setting electrons in action to create commercial electricity, the electrons of neuro-electricity (i.e., nerve energy) are set into increased action by cell oxidation. This means that electrons in the atoms which make up the cells are set in action by the oxidative process of the cells. On this point Crile has this to say: 'We conceive that the electricity liberated in the cells of the central nervous system by oxidation and conveyed to the cells (forming the different parts of the body) by the nerve fibres, act as a catalyzing agent by means of which the cells of the body are controlled by nerve action.'

"And again: 'We may consider that electricity keeps the "flame of life" burning in the cell, and that the flame (oxidation) supplies the "vital force" of the body, in accordance with this concept; therefore, the cell is an autonomic mechanism; life as we view it is an expression of the activity of this autonomic mechanism.' (Pages 195 and 15, *A Bipolar Theory of Living Processes*.)"

The Body: A Bipolar Mechanism

Next to the discovery of the body being an electrochemical mechanism, the discovery made by Burr, Crile, et al., that the body is also bipolar is the more significant, for it takes the two concepts to give a better understanding of the why of nerve flow. The body being bipolar, the brain is the negative and the peripheral tissues are the positive poles, according to the new interpretation of electrical phenomena developed by nuclear physics; and according to this same interpretation, the flow of nerve or neuroelectric energy is from negative to positive poles. This is the reverse of the old concept of electric flow, which was from positive to negative poles.

A further study of electricity and its behaviour gives a better basis for an understanding of the phenomena of nerve energy and of nerve flow. Having discussed nerve energy earlier in this article, we are now ready for the phenomenon of nerve flow.

Nerve energy is primarily a direct (DC) current, as in electricity—meaning it has a one direction flow, from negative to positive poles. In the body, as we know, the flow is from brain to periphery. This enables the brain to co-ordinate all body activities. However, this poses a problem. If neuroelectric energy has a one-direction flow from brain to periphery, how then is a demand for more action in a peripheral organ, gland or tissue going to reach the brain?

Reversed Polarity

In answering this important question, let us first consider the nature of a demand. In the chapter "Cell Responses and the Body's Lines of Defence", we stressed the point that cell groups (groups of organized cells) inherit a defence mechanism—a pattern of response—and that it goes into action only as or when something happens to change cell environment by creating a stress or distress in the cell groups (organs, glands, muscles, etc.). Thus food in the stomach, waste material in the bowels, fluid in the bladder, cause stress in these organs. On the other hand, an injury, body poisons or a diseased condition cause distress in the parts involved. These affect cell environment and thereby create demands to which the body responds. But when the neuroelectric flow is of the DC or one-direction-flow type from brain to periphery, again we ask: How is this demand going to reach the brain? The answer is that neuroelectric energy must reverse its flow and in order to do this, body polarity—as is involved in this situation—must become reversed. The area involved must become negative to the brain, at least temporarily. Such as this is precisely what happens and this is how it is done. The changed cell environment that such factors bring about, acts as a converter. This changes or converts the direct (DC) current into an alternating (AC) current, similar to the convertor in the electrical field. The AC is a type of current that flows in two directions—forward and backward. In the body, such a flow is from periphery to brain and brain to periphery, over afferent and efferent nerves.

Thus we have in the reversed polarity concept an explanation, brief to be sure, of the nature of a demand and of how the body responds to it. But such a concept raises many questions, questions on which we need answers. When they are answered, however, we shall have a still better understanding of chiropractic and the answers will be scientific, not philosophic—something badly needed. A few of the questions on which we need answers on the basis of this concept are: How and when is the AC changed back to the normal DC flow? Can a demand become sufficiently intensified to force the body to respond at such a high rate that its response gets out of control and goes wild? And can such a high rate of response bring about a polarization of such areas and cause the reversed polarity to become locked in reverse? The answers to these questions, as we have found in our research, will be the material for my next article, "Cells and Functions Gone Wild."

CELLS AND FUNCTIONS GONE WILD

It is the body's power of response which makes the Demand-Response relationship possible. This power, inherent in the body, has been recognized by most all doctors and researchers. It has variously been designated as being a "compensatory reaction", "the body's defence mechanism", "alarm reactions", "the body's adaptive process", "energy-producing reactions", "the power of self-preservation", or more simply "a patient's comeback". It has three types of manifestations as we have previously observed: instinctive action, urges and cell-group response. The latter of which is the more important factor to us as doctors. Without it, no curative process could take place and no person would ever get well. The cell-group response is an inherited pattern of response, in which the right thing at the right times is done or would be done if there were no interferences to the co-ordinative power of the body—the central nervous system. Its tendency is always towards normality which only nerve interference and injuries can change, on the one hand. On the other hand, it can be forced to function at such a high level that its response can and frequently does get out from under control.

The body does a pretty good job of keeping all its activities properly co-ordinated. It is only when the job becomes too great or nerve interference prevents the body's handling it adequately, that sickness develops. Were this not true, there would be much more sickness. The demand to which the body responds can be mechanical, chemical or emotional. Either of these can force the body to function at a higher level. Thus the body steps-up its activities to meet a demand. The question arises; can the body step-up its activities to an even higher level in order to meet a demand that persists and increases in intensity, and still remain normal? In a previous article—What is Disease?—we answered this question and the answer was, no. The integrative, co-ordinative force of the body is only constructive so long as it operates on a constructive level. Otherwise it become disintegrative and destructive. Thus the operative force of the body is much like fire, steam or gasoline. It is only beneficial so long as it is adequately controlled. The force which holds the body together, can figuratively cause it to fall apart when it gets out of control. A demand too great for the body to meet, brings this about, or it can occur through failure of a message to get through once the demand has been met, and so the excessive activity is still maintained to the degree that it becomes destructive.

How Control is Lost

As an illustration of loss of control through excessive demand—I can hit my hand hard enough to hurt it. Thus I create a demand to which the body responds by sending more nerve energy to that part. If nothing irritates it further, nature will soon correct the effect. But if I keep hitting it, the body will keep stepping-up its response to an ever higher level to meet the demand, until the point is reached where the response no longer can be controlled and serious trouble develops. Such as this can and does take place in cancer. It starts, not in abnormal but in normal cells—cells which are forced to respond to such a high demand that they cannot remain normal. Their function and growth go wild. Cancer develops. Thus any change in size or function of cells is but evidence of their having to adapt themselves to an abnormal environment, and the more abnormal the cell environment, the more abnormal the growth and function of the cells. Thus the process of destruction is started. Such is the effect of irritation which comes from mechanical, chemical or emotional sources. And the same thing is what takes place in most all ailments to a lesser degree. Cancer is only a stage in the development of disease, in which it has first gone through acute and chronic stages; for most all ailments, more especially those of the "itis" type can and frequently do turn into cancer. Thus, as we have found, all ailments under the surface are the same. Their names come from the tissues involved and the degree of the involvement. Their seriousness comes from their stage of development and their final stage is either CA or TB.

The foregoing paragraphs have outlined how control is lost through excessive demand. We are now ready to discuss the other way in which control can be lost. It is important for the body to slow down its stepped-up activities when the increased rate of activity is no longer required. But this slowing-down doesn't always happen. For afferent nerve interference can prevent the "cease fire" orders getting through once the battle against the diseased condition has been won, and it is allowed to continue its activity to the point that it becomes destructive. The following is a good illustration: If you hit a chicken on the head and knock it out, it will lie there inactive because the control lines are intact, but there is no intelligent control exercised. If, instead, you cut off its head—what happens? The control lines have been severed, so no control can be exercised over the activities of the chicken and they go wild. Such as that is a fair illustration of what happens when the "cease fire" order cannot get through and functions can no longer be controlled.

CARREL'S EXPERIMENT

Carrel's experiment with the chicken's heart, which he kept alive for twenty-eight years, is another good illustration. Some doctors have said that the heart was pickled. They are wrong. For it grew each day to the extent that it would approximately double its size each forty-eight hours and the excess growth had to be clipped off. Such as that couldn't have happened if the heart had been pickled. There would have been no growth, no indication of life. Besides, it would pulsate when stimulated by a micro-electric current. In order to keep the heart alive it was fed a certain diet, kept in the right sort of environment and its waste material taken care of.

If the excess growth hadn't been clipped off, it was estimated that in the twenty-eight years the heart was kept alive it would have grown to be the size of the world. And it doesn't take long to convince oneself that such as that could happen. All we need to do is to apply this story—A man said he would "shoe" another's horse if he would pay him a cent a nail and double the amount for each nail, at the rate that 2 times 2 are 4, 2 times 4 are 8, etc. It sounds simple and looks as if the owner of the horse would get the better part of the bargain. But would he? There are 32 nail holes in 4 horseshoes. Double each one of the figures until they all have been doubled 32 times and see what you will get. The final amount will be around 20 million dollars.

While the chicken's heart could live without any functional control, its size couldn't be maintained. A supply of the right sort of food could keep it alive structurally but could not control its functions or growth. So they went wild. When it is the *functions* that go wild, such ailments as these can develop; chorea, palsy, nervousness, tachycardia, dysentery, polyuria, acromegaly, tumours, etc. If it is the *cells* that go wild, these ailments can develop; TB, CA, pernicious anaemia, leukaemia, Hodgkins disease and other malignancies.

THE CORRECTION

It takes two different types of correction to overcome these two conditions. Through adjustments we can make it possible for the "cease fire" order to get through and have the stepped-up activity brought to an end once the battle against sickness has been won.

In order to overcome the other condition, namely excessive demands, it is necessary to overcome or get rid of the irritating factors which force the body into, and to maintain, an excessive response. This requires diet and good elimination. Both of which we have found to be important factors at Spears in the correction of any chronic ailment—factors of almost equal importance to the adjustment.

FOOD RESEARCH

When it was learned, largely because of the research of Burr at Yale University, that the body was an electrochemical organism, it opened up a vast new field of research into the electric phenomena of life, health and sickness. From then on, in order to utilize Burr's findings we needed only to learn how to create controlled demands and to measure the body's response to them—a problem that seven years' research has solved. Such is the basis of our research at Spears.

Another factor of a similar nature and one that has opened up an entirely new field of research, is one made by the writer (Wilson). It is this: foods are also electrochemical and their value as food can be measured electrically. So much so is this true, it can be said that food which has a low or no electrical reading will not sustain life; while foods which have a higher electrical reading have the higher percentage of food value.

FOOD AND BODY ENERGY

My first research into this question concerned foods from the standpoint of the contribution they make towards maintaining the body's functional activities. The previous concept **and** the one more generally accepted, was that carbonaceous foods—starches, sweets and fats—were the sources of body energy. In other words, these were the energy-giving foods. My research has proven this concept to be untenable. In the first place, if such as this were true, alcohol, which is almost pure carbon, would be the best food. Is it? Certainly not. Simply because when death takes place because of alcoholism, it is the result of starvation and not the alcohol. They starve to death because they keep so full of alcohol that they have little or no appetite or desire for food. They die in spite of the alcohol. Another myth about the value of carbon as a food, long since exploded, again concerns alcohol. It is now known that alcohol, instead of creating real energy, only gives a person a false feeling of being energized. It does so because it forces the body into a greater expenditure of energy in order to take care of or to get rid of the extra carbon. In other words, it is like a whip to a tired horse. It certainly doesn't give more real energy to the horse. It only helps to get rid of what little energy he has and he becomes weaker as a result.

When these facts about carbon were learned, why couldn't researchers go a step further and learn that all carbonaceous foods—foods having a carbon basis—do no more toward creating energy than does alcohol. Such is the information my research has revealed, but from another point of view. This point of view, as previously mentioned, concerns the electric measurements of the carbonaceous foods. While this holds true of carbonaceous foods generally speaking, it is true more in particular of such foods when they are refined, according to their degree of refinement. Thus according to their electric measurement, fats, oils, greases and alcohol register zero. Refined flour, sugar and honey register only 3 micronutrin units; whole wheat flour 24, potatoes 24, rice 12. Thus in terms of micronutrins or life elements these carbonaceous foods, even in their natural state, create little body energy.

The Value of Carbonaceous and Proteinaceous Foods

If the carbonaceous foods do not energize the body, of what value are they? Their value is fourfold: (1) Carbon is necessary in the trigger mechanism of the cells of the body, through which their spark of life is generated. (2) Carbon is the factor which creates body warmth, a factor that is as important as energy. (3) Carbon is the important factor in fevers, through which the body cleanses itself "as by fire". (4) It is from carbon that carbon dioxide is created, and carbon dioxide is necessary in the function of the vasoconstrictor and respiratory centres and as a factor in the ventilation of the lungs. Such is the value of the carbonaceous foods. As such they are necessary, not refined, but in their natural state; even though they create little or no energy.

Research before coming to Spears, as well as research in Spears, has proven that the real energizing foods are the proteins. In this we have found that the proteinaceous foods have two other important functions aside from their energizing effect. They maintain cell virility and the body's ability to function normally.

Cell Virility

The fighting power of the body comes from the level of cell life and the level of cell life is determined by body voltage, called reserve energy. It is a product of the cells which are micro-dynamos. Voltage is high in health—low, in sickness. When voltage drops and we cannot build it up again, such a patient can not be cured. Nor is he long for this world. Thus because the level of cell life which gives the cells their fighting power and virility, is so important in the cure of the sick, it is our No. 1 research problem. We must know the factor which builds it up. This has taken us into food research, and the important factor about foods in building up cell virility is the proteins. This is the order of their value: red lean meats, liver, eggs, cottage cheese, fish and chicken. Proteins from nuts, grains, legumes and other vegetables are as a rule, too complex for the weakened organs of the sick to properly digest and assimilate. When patients are too weak to handle meat, they are given soups and broths made from meat stocks.

The Body's Ability to Function

The body's functional ability, called functional energy, comes from its amperage. Amperage is voltage in action, drawn into action because of and in relationship to a demand. Thus, in order for the sick to recover, it is as necessary to have the required amperage as it is the required voltage. Obviously when voltage is high, amperage is seldom a serious factor. Its importance comes from the fact that when its level is high, it means the body is overworking, and when such as this happens it is because of an accumulation of poisons or because of pathology. When the level of amperage is low, the body can no longer maintain its functional activities normally. This is frequently followed by cell exhaustion which, if it is not stopped, leads to death. Here again, meat proteins have proven the best means of building up body amperage, as in body voltage.

Leafy Vegetables

It must be recognized that the values of the proteinaceous foods are best obtained when eaten with the green vegetables, both cooked and raw. And the more valuable of these, according to their micronutrin units, are the leafy vegetables which grow in the sunshine, on which the sun-rays have a more direct action—such as celery, lettuce, red beet tops, spinach, etc., which register from 81 to 106. Next come the vegetables on which the sun-rays do not have such a direct action, even though they grow in the sunshine. They are the legumes, the sprouts, the vine vegetables and the grains, which register from 17 to 68. The least valuable in micronutrin units are the tubers and the root vegetables. They register from 24 to 41. Fruits are likewise low in such units, in that they register from 16 to 47. The foods having a low unitage of measurement have other values, such as bulk and vitamins; but they do not mix so well with proteins as do the green leafy vegetables, so are not valuable in building up cell virility or the body's functional ability.

Food and Nerve Flow

Not the least important of food values comes from the contribution they make towards facilitating nerve flow, which comes from their mineral content; for without minerals there would be little or no flow of nerve energy. This point can best be understood by bringing to mind some experiments carried out some years ago, in which pigeons and dogs were fed demineralized foods. Within a couple of weeks on such a diet, they all became paralyzed. Thus it is obvious that minerals are necessary in nerve flow. This coincides with my findings, for it is the mineral content that I measure; which explains why it is that the more highly refined a food is, the less micronutrin units it registers and the less valuable it is as a food. Thus I have found a simple method of measuring food values electrically, through which such values may be easily and accurately determined.

It is the minerals in the nerves which make possible the transmission of nerve energy. Thus, to the extent that the mineral reserve of the body becomes depleted, to that extent will the nerves lose their ability to convey energy to the different parts of the body. And oddly enough, it is such patients as these which have a low body voltage and a low amperage; which means a low power of health response, as our research has revealed. All of which emphasizes the importance of a proper diet. However, the best type of diet for each patient, which the diet question resolves itself into, is best determined on the basis of the patient's rate of digestion as to which foods or food combinations speed up or slow down the rate of digestion. (Something which we have to discuss more fully in a future article, as we will also discuss the part that acids, alkalines and toxins play in causing interference to nerve flow; factors which do not come within the scope of this article.)

Dehydrated, powdered vegetables, grains, meats, fruits and distilled water were used in my tests; and the measurements were made by direct electrode contact. Powdered products and distilled water have proven to be the more scientific in making such experiments, for accurate measurements can only be made when the relationship of solids to liquids can be accurately controlled, which can not be done with foods in their natural state. We could have no basis for comparison of values between such foods which have a content of water ranging from 35 to 94%. But what is the value of powdered foods? When they are properly prepared, none of their food values appear to be lost, according to our findings.

THREE FACTORS OF HEALTH RESTORATION

PART I: CELL ENVIRONMENT

May I emphasize that this series of articles is not presented for the purpose of trying to prove or disprove anything; only to present the concepts developed from our research. And may I also say that already, as a result of our thirty-six discoveries and developments in the fields of etiology, diagnosis and health restoration, it will take from five to ten years to adequately explore, thoroughly test and completely integrate these factors. Therefore, all I can do in these articles is to present the high spots and to give an overall outline of the concepts developed as a result of our research. Such is true of the Three Basic Factors of Health Restoration.

As a result of a better understanding of the factors involved in health and sickness, I have been able to obtain a more comprehensive understanding of the factors involved in health restoration. And such factors, as we shall see, become a good yardstick through which to measure the efficacy of the methods employed in the different branches of the healing arts. These factors, I have found, can be classified under three headings. They are: Cell Environment, Cell Response, and Cell Co-ordination. Cell Environment is the subject of Part I.

It is a change for the worse in cell environment that brings sickness; a change for the better that brings health. Therefore, the state of the cell as to its shape and functions, is so important in sickness and in health that it could be said, "as the cell is, so is health."

Nerve Interference

In order, therefore, to understand the restorative process involved in health restoration, we must know the factors which change cell environment. In the past, Chiropractic has considered nerve pressure as being the only factor involved. The Wilson research shows that nerve pressure is only the initial step in the chain of causative sequence which brings about the changes that lead to sickness. As a result of the new information obtained, it is advocated that the term "nerve-pressure" be dropped and the term "interference of nerve flow" be substituted—in view of the many factors we have found to be involved in preventing a "directive intelligence" from keeping all activities of the body co-ordinated into a smooth-running organism. Part III of this series will discuss all phases of interference to nerve flow. In our discussion the new term "neuro-stasis" will be used; a term which better describes the effect of interference to the nerves, for it can be induced by a light as well as a heavy pressure. It is mentioned now as something to be thinking about. It opens up a whole new world of thought, in which we will see our science in a better perspective.

Next to a neurostasis, nutrition and elimination are the important factors in cell environment. Yet we must be mindful of the fact that the adverse effects of wrong nutrition and sluggish elimination on cell environment can be causative factors in, as well as end effects of a neurostasis; as we shall learn in Part III. Part II takes up other important phases of these factors.

Nutrition

Having discussed in my last article, "Food Research", the misconception of the foods which create body energy, including the importance of vigorous cells and the foods which best build up their virility, it leaves the rate of digestion as the next important dietetic factor needing to be discussed. In this we should keep in mind that no completely digested food ever hurts anyone; that the best food partially digested can be injurious, and that poor food more completely digested is more nourishing. Thus the rate of digestion—something that has largely been overlooked—is one of the most important factors in dietetics; for partially digested proteins create proteinoids and partially digested carbohydrates and hydrocarbons (fats) create carbonoids—both of which are important factors in creating an adverse cell environment. Therefore, in this article I shall discuss:

(1) The rate of digestion;

(2) The carbonoids and

(3) the proteinoids.

1. *The Rate of Digestion*

The main thing which slows down the rate of digestion is wrong food combinations, and the question of right combinations does not involve foods per se so much as they do the physical makeup of the patient—as to whether he has a slow or fast digestion. Nor is this question as complex as it would appear. It is easy to understand; for if a skinny patient had a fast digestion, he would be fat; if a fat patient had a slow digestion, he would be skinny. It is as simple as that. The fleshy person, having a good digestion, needs not to be so concerned over food combinations as over eating good nutritive foods. He is the one who can digest "nails" so does not have to worry about diet so long as his foods are nutritive. That is his main concern. How different with the skinny person! He must watch his combinations or become sick.

Needless to say, that more often the healthy person is the one who has a fast digestion; the sick person, a slow digestion. This is true with most all sick people, especially the chronically sick and those whose ailments have been greatly debilitating. Such as these usually have in addition to a slow digestion, a sluggish liver, sluggish elimination, hypotension, low red blood-cell count and are anaemic; which means a low level of cell activity, slow functional activity of cell-groups and a low power of health response. They are the ones we have difficulty in helping.

The factors which slow down their rate of digestion are: greases, sweets, starches and fruits, Yes, fruits! Eaten either as a meal or between meals, fruits have great dietary value. But not with proteinaceous or carbonaceous food, for the man who has a slow digestion. Green leafy vegetables are the neutral food. They can be eaten with proteins, starches or fruits. Thus potatoes and bread or fruit, which shouldn't be eaten together, can be eaten with green leafy vegetables, cooked or raw made into a salad, to which cottage cheese, cheese or avocado can be added to increase their food value. But proteins other than those just mentioned, should always be eaten with green leafy vegetables, cooked or raw;

preferably the latter, except for the seriously sick; and the proteins from lean red meat, liver and eggs are the best. The proteins of nuts, soyabeans, legumes and cereals are too complex for a patient with a slow digestion to handle. A hard-working man can handle them, but not a sick person. (Further factors about diet are explained in my book *A New Slant to Diet*.)

2. *Carbonoids*

While secretions from the pancreas and intestines aid in the digestion of carbonaceous foods, the saliva is the important factor in digestion before the food leaves the stomach. With these digestive factors working and each contributing its part, carbonaceous foods are properly digested. When they are thus digested they become carbon, from which the warmth of the body is largely maintained and, as carbon, they take part in the trigger mechanism of the cells, from which comes their "spark of life". Or they become carbon dioxide and perform useful functions in the vasoconstrictor and respiratory centres and in the ventilation of the lungs.

For the carbonaceous foods to serve such useful purposes, they must be completely digested. Before they reach the digestive stage of dialysis in which they are absorbed into the blood stream, they are eliminated through the bowels. Therefore it is not the first stage (before dialysis), or the last stage (carbon and carbon dioxide) of digestion in which we are now so concerned. It is with the middle stage, the stage in between dialysis, and carbon or carbon dioxide,—the food is in a condition in which it has left the carbonaceous stage and has not yet been digested sufficiently to be utilized by the body. It is neither food nor nourishment. The state of food in this stage of digestion is what I call "carbonoids". In such a state they are poisonous to the body, not only because they help clog the intercellular spaces but because they are also the factors which cause an excess acidity or alkalinity; both of which change cell environment adversely.

Carbonoids can create an overacidity or an alkalinity in the same person at different times. This is done as a result of the process the carbonoids are taken through in order to break them down to where they can be eliminated, which can not be done while they are in their carbonoidal state in the intercellular fluids of the body. To change them, they are taken through a process of electrolysis in which either positive or negative electrons are liberated.

If it is the former, there will be a hydrogen ionization which is an acid; if the latter, it will be a hydroxyl ionization which is an alkalinity. Why? We have found the answer in cell virility. When the cells have a low virility, the ionization will always be hydroxyl—an alkaline; when their virility is high, the ionization will always be hydrogen—an acid. This explains why an ailment is always overly acid at its onset; but becomes alkaline when it becomes chronic, according to its degree of chronicity; and why no chronically sick person passes away until he reaches a high degree of alkalinity, except from haemorrhages, a heart attack, etc.

3. *Proteinoids*

Unlike the carbonaceous foods, the proteinaceous are digested by hydrochloric acid (HCL). However, when fruits or starch or greasy foods are eaten with meat proteins, the demand for saliva offsets the demand for HCL. Then one neutralizes the other. The result is that neither type of food is digested completely. The outcome is carbonoids and proteinoids. As with the carbonoids, the proteinoids must go through several stages of digestion. Seven, to be exact. The fourth is the stage of dialysis. Up to this stage the partially digested proteins are eliminated through the bowels, so need not concern us. At the fourth stage the partially digested proteins can be absorbed into the blood stream but cannot be assimilated by the cells. To do so they must be completely digested, at which stage they become amino acids. So we do not have to be concerned with the last stage of digestion. It is the stage between dialysis and amino acids in which we are concerned. The state of food in the in-between stage of digestion is what I call proteinoids.

The proteinoids when in the blood stream and when circulating through the intercellular spaces, cannot be eliminated through the regular channels of elimination. The only other channel left is the mucus membrane. Then they are eliminated as mucus. The effect on the mucous membranes is one of irritation. As a result, proteinoids are the contributing factor in all of the so-called "itis" ailments. And inasmuch as the "itis" ailments, as our research has revealed, are the carcinogenic ailments—the ailments from which cancer more often develops —the proteinoids are, therefore, one of the most common factors in the development of cancer, simply because of the irritating effect the proteinoids have on the cells, which forces them to function at an excessive rate. Finally their rate of function become so high that it can no longer be controlled. The result is that the cells go wild. Such is cancer. All because of the proteinoids changing the cell environment excessively.

All ailments under the surface are the same and come from the same basic factors—mechanical, chemical or emotional trauma from injuries, body poisons or emotional upsets which change cell environment—and their names come only from the tissues involved, the degree or stage of the involvement. If this is true, and research is proving it to be, then the end stage of most all ailments is cancer. Such as that takes place more often when an ailment is allowed to become chronic and keeps flaring up. All because body poisons maintain an abnormal cell environment, of which the carbonoids and proteinoids are important offenders. In this, proper diet together with good elimination, are the answers.

ELIMINATION

The part that neurostasis plays in sluggish elimination will be discussed in Part III. More is known on the question of elimination than on other phases of body functions, so not too much need be said at this time. But it should be emphasized that so long as toxins are present in the body, elimination will be poor no matter how many times the bowels move each day. Besides nerve energy, good elimination requires bulk—plenty of it—and bowel training. Bulk comes from the leafy vegetables and whole grains. Training comes from establishing regular times for daily bowel relief. Too often through long neglect, it takes considerable time and patience, along with good adjustments, in order to get the bowels re-trained and re-activated. (The emotional factors involved in elimination and in creating a bad cell environment, and the involvement of emotional factors in most all ailments, will be discussed in a separate article at the conclusion of this series. This question is important enough to warrant special consideration.)

ADDENDUM

It will no doubt be recognized that the doctor who does nothing but give colonics or uses other body-cleansing methods, and the doctor who relies entirely on diets, will obtain pretty good results. They will be re-establishing normal cell environment, at least to a degree. While such methods are good so far as they go, they are not the complete solution in the problem of health restoration, as we shall see.

PART II. CELL RESPONSE

The continued existence of the cells is made possible by their inherited ability to take in nutriment and oxygen, get rid of their excreta, create new cells and to defend themselves. Such as these are things the cells must meet successfully or cease to exist. Thus it is the inherited power of response of the cells and their ability to meet such demands, that makes life and its continued existence possible. Such are the inherited response patterns of individual cells. The response patterns of the groups or organized cells—the body as a whole or of its organs, glands or other parts—are something else. These inherit group-patterns of response or their existence could not be maintained. Such patterns are called instincts and urges. They involve the total organism. But there are other patterns of response which do not involve the total organism. Instead they involve the organs, glands, muscles, ligaments, bones, fasciae, membranes, etc., and enable them to meet the demand made on them individually. It is with this latter type of response, the response of organized cell-groups, that we are now concerned.

We frequently hear doctors say, no matter their system of healing: I use this, that, or something else, or I use this therapy or that therapy, because it stimulates action and aids in the correction of an ailment. Does it stimulate action? Can the body or any of its parts be stimulated? Or is there such a thing as a stimulant? The answer to each of these questions is, strange as it may seem, an emphatic, NO. Simply because there is no such thing as a stimulant. Such an idea is one of the archaisms of the past which is still used without stopping to learn if it is really true. The idea was developed at a time when little or nothing was known of how or why the body functions. No, it is not what a doctor does which stimulates action; it is what the body does about it. Little has the doctor realized that all he can do is to create a demand and elicit a more energic response. Thus, to use an illustration on something we all know, it isn't the physic which moves the bowels but the bowels which moves the physic. The physic only creates a demand to which the body responds, and in its efforts to get rid of the physic, it moves the contents of the bowels as well. This is one of the archaic ideas of the past that we can do well to get rid of. It is so misleading. In this respect we should do as Brookes emphasized when he said: "If a thousand old beliefs are ruined in our march to truth, we must still march on."

There are factors, however, which appear to have a stimulating effect on the body. For instance, if you happen to like candy or beefsteaks, your mouth will "water" or your stomach increase its secretions when you think of one or the other. Such as that happens, we are told, because of the stimulating effect those things have on the glands of the mouth or the stomach. But such responses are conditioned, not natural heritages; for children are not born with natural appetites for candy or beefsteaks. Such appetites are acquired, and their learning to like such things, says psychology, is a process of conditioning. Thus the "watering" of the mouth and the stepping up of secretions in the stomach are nothing but conditioned responses.

FACTORS WHICH CREATE DEMANDS

No matter the system of health restoration a doctor uses, it will come under one or more of these headings. But what the doctor hasn't learned is that each of these methods but creates demands, to which the body responds by stepping up its activities. Such methods are: general adjusting, general massage, manipulations, medicines, modalities, counter-irritations, hot and cold packs, hot and cold baths, hyperpyrexia (fever therapy), hypopyrexia, ethylchloride hydrothermal revolution, skin brushing, scalp massage, drinking a lot of hot or cold water, traction, muscle re-education, cranial remoulding, nerve and cell goading.

It may appear strange to see general adjusting included in the list. It is included because, aside from the relaxation it brings about, it creates demands and elicits more energic responses; but effects no specific corrections. What is meant by a specific correction is that one place in the spine has more to do with a patient's ailment than do all others. Unless this one place is properly corrected, the best and the more lasting results are not being obtained.

CREATING EXCESSIVE DEMANDS

In the past the idea was quite generally accepted, and such is generally true today; that in order to effect a cure, all we need to do is to step-up the curative action of the body. Is that really our problem? No! It is relatively simple to do just that; for it isn't difficult, as a rule, to step up the curative process. So that isn't our problem. Our problem is to build a body structure that can *stand and sustain the accelerated activity necessary to restore health*. On the point of what the curative process actually is, as our research is revealing, and what the ideas of it are, are usually so much at variance, that if there is any resemblance between the two, it is purely coincidental. A point not generally considered in stepping-up the curative action of the body, is cell exhaustion, and such as this is a factor in all chronic ailments according to their degree of chronicity and debility. In such ailments the cells of the body, from their long battle against the ailment, become tired and worn out, and too often have reached a state of partial exhaustion. In such a state care must be taken not to step-up the curative process too rapidly. To do so may bring on complete cell exhaustion and a rapid death.

Such a conclusion has been obtained from a study of several hundred cases, of which this case is quite representative. His body voltage which represents the level of cell life, had been no more than 50 for around two weeks. Then, within two days it jumped to 83 points. Such an increase would appear on the surface to be good; for it would appear that we were winning the battle against his ailment, by building up the level of cell life. But experience has taught otherwise. In the first place, at fifty points, it was hardly possible to save him or to build up the level of cell-life maturity. He was too far gone. But even if it could be built up, which we are learning how to do, it wouldn't have been built up so fast. It would have been a slower process. Thus, for it to jump thirty-three points within two days, we knew, was dangerous. We knew the cells of his body could not stand such a strain. Nor did they. He passed away within eight days. On all cases on which we have found such a rapid acceleration, death has taken place within two to fourteen days, without exceptions.

We are well on our way to solving the problem of cell exhaustion which is our Number One research problem. When we do solve it we shall be a long way toward solving the problem of chronic ailments as well as cancer. When we can prevent or largely overcome cell exhaustion, we will not have to worry over stepping-up the curative action of the body too rapidly—something which takes place many times in the chronically sick, where there has been a great debility. At Spears, where we are testing many instruments and methods in our research department in our efforts to find the best methods of stepping-up cell activity, aside from building up cell virility through a better diet, we have found that cell and nerve goading, as developed by Dr. Leo Spears, is the more effectual. Through its use we can create demands and elicit more energic responses, which is proving to be the best aid to adjustments in bringing about more rapid cures.

CONCLUSION

If in our work we do no more than to create demands and step up the body's response, all we are doing is bringing relief. In which case we are doing no more for our patients than are the other doctors. Thus if we only change cell environment through diet and/or establishing better elimination, or if we only step up the curative process through creating demands and eliciting more energetic the curative process through creating demands and eliciting more energic responses, we are only obtaining temporary results. To obtain permanent results, it requires another factor which will be discussed in Part III under heading of Cell Co-ordination.

PART III. CELL CO-ORDINATION

No doubt you have heard this old statement. "He drew a circle that left me out; I drew a larger circle that took him in." Edwin Markham's statement has an application in our present discussion of the three basic factors of health restoration. The allopaths, homoeopaths, osteopaths, naturopaths, chiropractors, Christian scientists, faith healers, etc., each has drawn a little circle which keeps all others out. Now, from my study of the biodynamics of health and sickness, I have drawn a bigger circle that has taken all of them in; for that is precisely what the last three articles of this series has done. It has classified all health methods of the healing arts into three general divisions and put them under one general heading, the one that heads Parts I to III of this series.

The reason for the classification is simply this, based upon our research findings; that no matter the health system used, there are only three general ways in which sick people can be benefited; either by improving

(1) Cell environment; by eliciting a more energic

(2) Cell response by creating controlled demands which accelerate cellular activities; or by

(3) Cell co-ordination—the co-ordination of all the cell groups of the body into a smooth-running organism, which means health.

Cell Co-ordination

Before we can enter into the discussion proper of cell co-ordination, we need to answer two questions. Does the body need a co-ordinating power? If so, is such co-ordination chemically or neuro-electrically controlled? The answer to the first question is an emphatic, YES. For the body could no more exist without a system of control, than could civilization exist without law and order. Therefore, concurrently with the evolution of life, there had also to be evolved a system of cellular and functional co-ordination, or the higher forms of life could never have come into existence. The answer to the second question centres on the basic difference between medicine and chiropractic, which resolves itself into a question of whether the body is chemically or neuro-electrically controlled. If the body is chemically controlled, the medical concept is right; if neuro-electrically controlled, the chiropractic concept is right. What are some of the basic facts back of each concept? Here are some of the answers to the question.

Chemical Control

Evidence points to the fact that early forms of life were chemically controlled. To quote Butler: "Primitive methods of regulating and controlling body activities were chemical, since the simple primitive forms of life had no nervous system." (Man Is A Microcosm.) To which we can add the research of Carrel, the Ralson Club and the work of tissue culture, through which pieces of chicken's hearts, tendons, muscles, nerves and bones were and are kept alive for indefinite periods of time. In the case of tissue culture they are kept alive for surgical purposes. And to this can also be added the work of Cannon of Yale University, who cut all the nerve connections between the visceral organs and the central nervous system of cats and they continued to live, but were reduced to a mass of protoplasm rendered totally incapable of caring for themselves. Such as these were chemically controlled.

Neuro-Electric Control

To again quote Butler: "In the higher forms of life this method of (chemical) control has been supplemented by a much quicker method—the use of nerves to send signals from one part of the body to another, but the older methods of control ... still underlie the whole nervous system." To quote MacLeod's *Physiology in Medicine*, 9th ed.: "Some superficial responses (of the body) are but skin reflexes. Medium depth responses involve the immediate nerve reflexes. While a response which depends upon more factors, through which to effect curative changes, involve the co-ordinating effect of the cerebrospinal system." And to quote *Gray's Anatomy:* "Broadly stated, the nerve system connects the various parts of the body with one another and co-ordinates them into one harmonious whole in order to carry on the bodily functions methodically."

In the *Textbook of Physiology* the explanation of the difference between chemical and neuro-electrical control of the body is more explicitly stated under the heading of "Neural Versus Hormonal Control." To quote: "The two systems of intercommunication between the various parts of the body, namely, the natural and the hormonic, differ in two or three important respects; the action mediated by the nervous system is generally localized, specific, and speedily executed; that following the production of a hormone is vastly slower, frequently more widespread throughout the body and, because of the continued presence of the hormone, of a more lasting nature." The following are some of the main factors to which he refers: The influence of the pituitary secretions on the size of the body; of the thyroid on the emotions, and of the sex glands on the contours of the body, etc., while all speedy activities, including the functional control of all organs and glands of the body, are under the control of the system of nerves.

Another Basic Difference

Another basic difference between the concepts of chiropractic and medicine is this; does the electric energy found in the cells come from the outside or is it a product of the cells? If the energy enters the cells from the outside, again the medical concept would be right; but if the energy is a product of the cells, then the chiropractic concept is right. What are the facts? The answers are simple. All that is needed in order to answer those questions, is to recall Carrel's experiment with the chicken's heart which he kept alive for twenty-eight years; to recall the experiments of the Ralston Club in which parts of chickens' hearts, pieces of bones and flesh were kept alive for many years; and to recall tissue culture in which pieces of nerves, bones, tendons, etc., are kept alive for surgical purposes for indefinite periods of time. *None of these had outside sources of energy.* To keep them alive it was only necessary to keep them in a warm saline solution, give them the proper nourishment and get rid of their metabolic waste material. Thus they had their own source of energy.

What is the source of this energy? *It comes from the oxidative process of the cells which are micro-dynamos. In other words, the cells of the body create their own energy*—sufficient to take care of their basic needs for nutrition, elimination, protection and reproduction. And as long as the cells are alive and have these factors adequately cared for to keep them alive, the life of the tissues of the body into which they have become organized can be maintained indefinitely. Thus the chiropractic concept alone is tenable. The cells are the source of their own energy, except the cells of the brain. They are also the source of the nerve energy which co-ordinates all body functions into a smooth-running organism. Therefore, the full and complete co-ordination of body functions—which is the aim of chiropractic—is essential to life as it is also the basic fundamental of health restoration.

CO-ORDINATION OF BODY FUNCTIONS

Cell co-ordination means the co-ordination of body functions, for all parts of the body are only groups of organized cells; organized cell-groups, as we call them. The complex organization of cell-groups such as make up all higher forms of life, was only made possible (as previously mentioned) by a highly integrated nervous system. This strengthens the concept that the human body is nothing more than a product of cellular and functional co-ordination, co-ordinated to a nicety. Without such a co-ordination the body would be nothing —nothing but a mass of disorganized protoplasm. We can therefore accept this conclusion: inasmuch as a complete co-ordination of the cells and functions of the body has made it all that it is physically and mentally, a loss of such a co-ordination if carried on long enough, can strip the body of all that co-ordination has given it—health, strength and mentality—and, in time, can return it to a mass of unorganized protoplasm. Therefore human life is a product of completely co-ordinated body functions; sickness is a disruption of such co-ordination; death is a complete cessation of co-ordinated functions.

It should now be obvious why cell co-ordination takes its place by the side of cell environment and cell response, as one of the three basic factors in health restoration. This being true, it raises the question of the factors which disrupt co-ordination and when co-ordination is disrupted, and of the factors which re-establish complete co-ordination which leads to health.

DISRUPTION OF CO-ORDINATED FUNCTIONS

Long ago I dropped the use of the term "subluxation" and used "nerve pressure" instead, only to later drop this term for "interference to nerve flow". Now this term is also being discarded in favour of a still better term—"neuro-stasis.". Why? Because all previous terms failed to convey the real meaning of, or all the factors involved in, a disruption of the body's functional co-ordination. While the basic concept of chiropractic—that all functional activities of the body must be adequately co-ordinated if life and health are to be maintained—is even more unassailable than ever and has stood the test of time, we have failed to grasp the full meaning of the factors which disrupt functional co-ordination. It has taken time, experience and considerable research to obtain a better understanding of these factors. Such is the process of growing up and it is time chiropractic took off the swaddling clothes it was garbed in at the time of its birth. The concept of only osseous nerve interference taking place, while basically true, needs to be re-evaluated.

Instead of bony pressure being the only factor which disrupts functional co-ordination, research has revealed twenty-five factors. They come under four headings—mechanical, chemical, emotional, and neuro-electrical, as follows:

Mechanical	*Chemical*
Subluxations	De-mineralized food
Swollen nerves	Toxicity
Nerves constricted by their own sheaths	Excess acidity
Swollen tissues which constrict nerves	Alkalinity
Scar tissue	
Tumours	*Emotional*
Thin discs	Psychic sore spots and emotional
Herniated discs	undertones which cause muscular
Spinal distortion	and visceral tensions.
Necroses	
Congenital deformities	*Neuro-Electrical*
Bony spurs	
Calcium deposits	Neutralized polarity
Calcification of spinal joints	Reversed polarity
Edema (fluid accumulations)	Polarization
Tight clothing	

A comprehensive system of healing will take all of the above factors into consideration. In each patient, usually several of these factors are present as primary or contributing causes of illness.

NEGATIVE GALVANISM AS UTILIZED IN THE KEESEY TREATMENT FOR HAEMORRHOIDS

In the obliteration of internal or mixed haemorrhoids with negative galvanism, we have an application of artificial electrical energy to organic tissue to produce a therapeutic, curative effect which has many advantages over the two other commonly used methods of treating haemorrhoids—namely the surgical and injection methods.

Galvanism is direct current applied through two electrodes, one of them positive and the other negative. In negative galvanism, a small electrode is applied to the affected area, while a larger electrode is applied to another area of the body.

The galvanic method was first employed in 1867. It was later sponsored in a book published in 1892 by Baker—*A Treatment of Haemorrhoids by Electricity*. However, it was not until certain refinements and improvements in technique had been made by Dr. Wilbur E. Keesey, M.D. of Chicago, that the method came into wider use. One of Dr. Keesey's principle papers on the subject was published in the *Archives of Physical Therapy, X-ray and Radium* for September 1934, entitled "Obliteration of Haemorrhoids with Negative Galvanism".

C. M. Haynes, in his book *Electrotherapeutics*, published by W. T. Keener Co., Chicago, 1896, page 163, states:

> "Althaus made microscopical observations of changes in animal structures due to the electrolytic action of the negative galvanic needle. He found that the tissues were markedly contracted, and that there was neither inflammation, suppuration nor sloughing. When the current was applied to the blood vessels they became changed into solid strings due to disintegration of the blood and deposition of lamellated fibrin. He concluded that no animal tissue can withstand the disintegrating effect of the negative pole; that the force and rapidity with which disintegration is brought about are directly proportional to the strength of the current and to the softness and vascularity of the structures; and that the current could be safely and successfully applied to contract and disintegrate tissue, and obliterate blood vessels for surgical purposes."

When applied to haemorrhoids, the negative pole produces first a hydrolytic decomposition and then a contraction of the tissues. J. C. Webb, in his article "Treatment of haemorrhoids by electrolysis", *British Medical Journal*, 1, 457, March 1921—states that electrolytic destruction—is highly significant. Actual obliteration of the thrombosed mass is accomplished in one of two ways: it either absorbs as occurs in any simple contusion; or if a large, thin-walled haemorrhoid is treated, it ruptures, causing a discharge of the thrombosed elements into the rectum. Following this there is contraction of the underlying tissue with freedom from haemorrhage, absence of pain, and rapid healing of the parts.

Obliteration with negative galvanism is a far milder process than the electro-coagulation process; the latter is likely to produce the same post-operative complications as surgery—namely pain, sloughing, and haemorrhage. These complications are avoided with negative galvanism—and best of all, the strictures and recurrence of the haemorrhoids which so often follow surgical removal, are also avoided when negative galvanism is used.

Using the special electrodes and technique devised by Dr. Keesey, most cases of internal or mixed haemorrhoids, including those that protrude, can be cleared up with three to ten office treatments of a duration approximating ten to fifteen minutes each. No hospitalization or absence from work is required, such as occurs when surgery is used. The negative galvanic method is not applicable to cases of external haemorrhoids, however, the internal and mixed types which yield so well to the negative galvanism are said by Dr. Keesey to comprise 90% of all haemorrhoid cases seeking treatment.

When comparing negative galvanism with the injection methods for treating haemorrhoids, one learns that the injections are made with harsh solutions which cause contraction through scarification of tissue. The damage done by these chemical solutions is extensive, resulting in a contracted and distorted mucus membrane which loses its elasticity and becomes hardened, thus interfering with normal functioning in the future. Also, recurrences not infrequently occur after the injection method, particularly if an insufficient quantity of solution is injected. When using negative galvanism the action of the treatment upon the haemorrhoidal mass is kept under observation through a speculum, so that the intensity, duration and number of treatments can be adequately determined.

Negative galvanism does not produce the inflammatory, scarifying effect of the injection method, because the galvanic current acts upon the liquid content of the haemorrhoidal mass instead of on the tumour wall. One of the greatest advantages of the galvanic method is the resultant normal resiliency of the mucus membrane after the haemorrhoids have been obliterated.

The negative-galvanic method, although brought to its present state of development by a medical physician, and possessing undoubted advantages over other methods, has to date been adopted by only a minority of medical proctologists. To meet the public demand for the use of this method, the Keesey techniques have been imparted to a considerable number of chiropractic proctologists, who now provide large numbers of people with this service.

The subject provides an interesting example of application of a type of electrical energy to diseased organic tissue to produce a curative result without the unfavourable complications which attend the use of surgery or chemical solutions applied to the same condition.

Q

ENDOGENOUS ENDOCRINOPATHY

INDEX

ENDOGENOUS ENDOCRINOPATHY

INTRODUCTION

ENDOCRINOPATHY is a shortened term for endocrine (glandular) therapy. *En*dogenous means that the patient's own glands are stimulated to resume their proper functioning, as contrasted to *Ex*ogenous therapy in which animal gland extracts or synthetic imitations are administered to the patient.

Endogenous endocrinopathy, the method developed by the distinguished Dutch physician Dr. Jules Samuels, is of interest because it applies radio (short-wave) energy to the body in a manner that successfully treats diabetes, gastric and duodenal ulcers, multiple sclerosis in early stages, and other diseases—and its most spectacular results are obtained in cancer cases. In first and second stage cancer, case reports indicate that the cancer process is stopped in three to four months by means of this treatment; in third-stage cancer, prolonged use of the treatment saves about 80% of the cases, eliminates pain, and restores the patient to a much better state of health, even if the cancer is not completely eliminated in the cases which were far advanced before the Samuels treatment is begun.

This is a method developed by a medical physician and by 1951 it was being used by over fifty medical physicians in all parts of the world. A wealth of case data is available, also two books on the subject—both published in Europe. *Endogenous Endocrinopathy* by Dr. Jules Samuels, and *A New Light on Cancer*, by Dr. Francis P. de Caux, London, 1951.

The following information regarding endogenous endocrinopathy is summarized from these two books.

Endogenous endocrinopathy involves:

I. A new type of blood analysis, which resulted in the discovery of the Samuels endocrine concepts. This type of blood analysis performs three purposes:

   (*a*) Proves the validity of the new concepts.
   (*b*) Provides an accurate and scientific diagnostic method.
   (*c*) Provides continuous guidance for regulating treatment dosage, and for delineating the degree of progress attained.

II. A new set of concepts regarding:
   (*a*) The inter-relationships between various endocrine secretions.
   (*b*) The effects of endocrine functioning, upon cell functions of the body.

III.   A new method of using short-wave therapy in the treatment of cancer
and other serious ailments, in accordance with the Samuels concepts and
under the guidance of the new type of blood analysis.

I. A NEW TYPE OF BLOOD ANALYSIS

1. *Oxygen and Haemoglobin*

When the blood passes through the lungs, the haemoglobin (Hb) of the red
blood corpuscles absorbs oxygen (O), from the air breathed in, and is transformed into oxyhaemoglobin (HbO_2). During normal circulation through the
body, this oxygen is given up through the capillary walls to the tissue cells. The
time taken for haemoglobin to give up its oxygen to the surrounding tissue cells,
is termed the "reduction time". If this reduction time can be measured, it
provides an index of the respiration rate of the cell tissues of the body—namely
the rate at which they are able to absorb oxygen. This gives an insight into the
condition of cell metabolism which is more precise than any other analytical
method discovered to date.

Vierport, as far back as 1876, developed a method of examining the rate at
which oxyhaemoglobin gives up its oxygen in a piece of tissue in which the circulation is temporarily stopped by means of pressure. Hr. Henri Dausset of Paris
conducted research in 1935 into Vierport's method. Associated with him at that
time was Dr. Jules Samuels of Holland. Dr. Samuels continued the research
upon his return to Holland, and developed the type of spectroscopic examination which now makes the test clinically practicable for any physician to perform, if he has the required equipment.

2. *The Test of Reduction-time*

When making this test, the blood circulation in the web of skin between
thumb and forefinger is temporarily stopped by means of a clamp; a white light
is sent through the tissue and the transmitted light (the portion which has shone
through the skin and flesh) is examined with a spectroscope. In the yellow-green
part of the spectrum will be seen two black absorption lines. This is the spectrum
of HbO_2—oxyhaemoglobin. As one watches it, these two lines are seen to fuse
together. This is the spectrum of Hb and shows that the HbO_2 has given up its
oxygen into the surrounding tissue cells and lymph. The interval required for
the absorption of oxygen is measured in seconds, and is less than half a minute.
This test provides the key to the Samuels method.

3. *The Use of the Reduction-time Test*

All individuals in good health show the same interval or duration of *reduction
time*. All individuals having ailments induced by glandular imbalance, show a
reduction time deviating from the normal. Cancer, multiple sclerosis, and gastric
and duodenal ulcers have been shown by Dr. Samuels to have glandular imbalance as an originating factor, and to be curable through restoration of

glandular balance. The same applies to many other ailments not formerly classified as glandular. Dr. Henri Dausset was the one who discovered that the reduction time (the interval required for the cells to absorb oxygen from the haemoglobin in the blood) could be varied by stimulation of certain endocrine glands with certain short-wave frequencies.

This proved of great importance for both diagnosis and treatment. To complete the diagnosis, the pituitary, thyroid, and sex glands are given test irradiations with short-wave, and the reduction time noted before and after each of the three irradiations. The resulting graph shows the type of glandular imbalance for each particular case. The type of glandular imbalance shown by these tests, provides the information necessary to the selection of the appropriate treatment procedure.

The greater the respiration rate of the cells, the smaller is the reduction time interval, since the oxygen is more quickly exhausted from a given quantity of blood. Irradiation of the thyroid, stimulates the output of thyroxine, which increases the respiratory rate of the cells and decreases the reduction time. Irradiation of the sex glands has the opposite effect.

II. GLANDULAR INTER-ACTIONS AND THEIR RELATIONSHIP TO CELL FUNCTIONS

1. *Composition and Secretions of the Anterior Pituitary*

The anterior part of the pituitary gland is known to be composed of three different types of cells, termed the basophilis, the eosinophils, and chromophobes. It is generally accepted that the chromophobes do not produce hormones. From examination of clinical data and research on thousands of cases, Dr. Samuels came to the conclusion that there are just two types of hormones secreted by the anterior pituitary: the basophil cells produce the hormone which stimulates the gonads (sex glands) and is termed the gonadotropic hormone; the eosinophil cells produce the hormone which stimulates the thyroid, and is termed the thyrotropic hormone.

2. *Effects of the Anterior Pituitary Secretions*

The thyrotropic hormone accelerates cell functions, while the gonadotropic hormone retards or inhibits cell functions. When the output of these two anterior pituitary is properly balanced, cell functions are normal. The thyrotropic hormone stimulates thyroid action, the production of adrenalin by the suprarenal medulla, the production of testosterone in men and the luteohormone in women, and in conjunction with the thyroid hormone it activates and accelerates all cell functions, especially the dividing process of body and egg cells.

The distinction between the thyrotropic and the thyroid hormones should be kept in mind. Thyrotropic is the hormone secreted by the anterior pituitary's eosinophil cells, that stimulates the thyroid gland. The thyroid hormone is the one which is secreted by the thyroid gland itself.

The gonodotropic hormone stimulates the production of the follicular hormone in sex glands, the production of cortin by the supra-renal cortex, and while it *initiates* the process of dividing the body cells and the egg cells, yet it also controls and inhibits excessive division and growth of these cells.

The foregoing paragraphs contain some of the principal interactions of the glandular system; more details of endocrine functioning are shown in the appendix at the end of this section.

In the Samuels concept, the controlling glands are the pituitary, thyroid, and gonads. It is considered that the other glands support or reinforce the secretions of the three glands named.

3. *Effects of Glandular Imbalance*

(*a*) *Pituitary Imbalance.* If slight, can induce fatigue, headache, irritability insomnia, constipation, hyper-acidity, and general ill health.

(*b*) *Effect of Pituitary Imbalance on Other Glands.* Dr. Samuels states, and there is much supporting medical research to corroborate, that many diseases of endocrine glands are directly attributable to an imbalanced pituitary function. An actual or relative deficiency or excess of either the thyrotropic or the gonadotropic pituitary hormone, acting directly on the thyroid or gonads, can cause a corresponding disfunction in the gland concerned.

Under-activity of the thyroid may arise either from a deficiency of the (pituitary) thyrotropic hormone or from an excess of the (pituitary) gonadotropic hormone. The symptoms are identical in either case. Only the Samuels diagnostic method permits the differentiation of the cause. Under-activity of the gonads (sex glands) may likewise result from a deficiency of the pituitary gonadotropic hormone or from an excess of the pituitary thyrotropic hormone, the latter causing a relative suppression of the gonadotropic centre. These two conditions give rise to a whole series of disturbances including infantilism, excessive bleeding during menstruation, etc. Analogous disturbances also occur in the other endocrine glands, for each gland is activated by one side of the thyrotropic-gonadotropic balance, and inhibited by the other side.

(*c*) *Effect of Pituitary Imbalance on Tissues and Organs.* In his discussion of "hormonal damaging diseases", Dr. Samuels and Dr. de Caux show the serious consequences of a prolonged secretion in the body of an excessive quantity of a hormone. The constant action of an excess of one or both anterior pituitary hormones, working either directly through the blood, or indirectly through specific nerve tracks, can cause damage to peripheral organs, tissue systems or glands. They found that numerous diseases of the internal organs such as heart, lungs, liver, stomach, pancreas, kidneys, etc. fall within this group. The excess of hormone secretion may damage the tissues of organs and furthermore, the tissues of systems such as connective, neuroglia, joints, etc. and is also responsible for numerous affections of vital functions, such as fat, water, pigment, temperature, etc.

Formerly, all these diseases have been treated merely symptomatically by medical physicians, as the glandular cause was not known.

(*d*) It is Dr. Samuels' conclusion that the actual localization of a disease depends largely on predisposition—innate inferior construction or acquired pathological change of the organ or area concerned. The combination of what he terms the central cause—an over-active pituitary function, with the local predisposition, results in a disease in a specific organ. In the stomach, various disturbances may be caused, one of the most common being the stomach ulcer; damage localized in the joints may result in arthrosis or as rheumatoid arthritis and arthritis deformans. This is becoming better known with the recent experiments in the treatment of arthritis with hormonal substances such as cortisone and ACTH.

(*e*) *Diabetes*. There has been a considerable amount of research by others, showing a close connection between the pituitary and the pancreatic disturbances which give rise to diabetes (a complete bibliography is contained in Dr. Samuels book, *Endogenous Endocrinopathy*). Dr. Samuels has been able to determine spectroscopically that about 85% of cases of diabetes are caused by an excess of the gonadotropic hormone, with 15% caused by an excess of thyrotropic hormone. The specific insulin producing cells in the islets of Langherhans are damaged by this excess of pituitary hormone, both directly by the bloodstream and indirectly via the vegetative nerve centre of the pancreas.

4. *Proliferation Diseases*

Dr. Samuels finds that this is an important group of diseases arising from pituitary imbalance, and includes benign tumours, malignant tumours (cancer), Hodgkins disease, and other more rare variations. One of Dr. Samuels' most outstanding discoveries arose from his spectroscopic findings that:

(*a*) All patients suffering from malignant tumours showed an over-active *thyrotropic* centre of the anterior pituitary, and

(*b*) All patients suffering from benign tumours showed an overactive *gonadotropic* centre of the anterior pituitary.

This pointed to a direct causal relationship between pituitary malfunction and tumour growth. One of the many proofs of the validity of this conclusion is contained in the fact that the injection into tumour cases, of pituitary extracts in which the *gonadotropic* hormone predominated, resulted in checking the growth of malignant tumours and promoting the growth of benign tumours, whereas just the opposite effects occurred when injecting pituitary extracts in which the *thyrotropic* hormone predominated. In this latter case, the growth of malignant tumours was promoted and the growth of benign tumours checked.

One of the additional proofs has been found by a number of investigators by autopsies on patients who died of cancer—showing enlarged pituitary glands and an increase in the number of eosinophil cells—the cells which are considered to secrete the thyrotropic hormone; this is the hormone which by injection is found to accelerate the growth of cancer. Research has been done by others in the past showing a connection between the pituitary and cancer; others showed a connection between the thyroid and cancer, while still others performed experiments showing a connection between the gonads and cancer, and in fact gonadal hormones have attained some recognition for use in treatment of cancer. But it remained for Dr. Jules Samuels to show the mutual inter-relationships between the pituitary, thyroid and gonads, which when unbalanced, lead to the train of reactions that appear to be a causal factor in the development of cancer.

Removal of a tumour by surgery or other destructive means, does not eliminate the glandular imbalance which led to the production of the tumour; one of the reasons for the strong tendency for cancer to recur after the use of the conventional medical destructive methods.

III. THE SAMUELS SHORT-WAVE THERAPY

Short-wave therapy has been in use for many years, but it never came close to producing the outstanding cures of serious ailments that are now being accomplished when used in accordance with the Samuels method. The secret lies in the method of application and the locations in the body to which the short-wave current is applied.

1. *Selection of Treatment Procedure*

The test irradiations of the pituitary, thyroid, and gonads referred to in Section I, with reduction time noted before and after each of the three tests, provides the mathematical data showing which of the three glands are underactive and which, if any, are overactive; also whether the pituitary centre deviating from normal is the thyrotropic or the gonadotropic.

In cases where pituitary function is *underactive*, the pituitary is treated alternately with the gland which *reinforces* the particular pituitary centre which is under-active.

Thus, if the gonadotropic centre is under-active, the treatment is alternated between the pituitary and the gonads. If it is the thyrotropic centre which is under-active, then the treatment is alternated between the pituitary and the thyroid.

In cases where the pituitary function is *overactive*, the pituitary is treated alternately with the gland which *inhibits* the particular pituitary centre which is over-active.

Thus, if the gonadotropic centre is overactive, the treatment is alternated between the pituitary and the thyroid. If it is the thyrotropic centre which is overactive, then the treatment is alternated between the pituitary and the gonads.

This is a completely new and revolutionary method of treatment, derived from the new understanding of glandular inter-relationships and functions resulting from the Samuels research and spectroscopic blood analysis. The treatment is aimed at the restoration of normal cell functioning in the body, through the elimination of glandular imbalance.

The wavelengths of irradiation found beneficial in the Samuels therapy, range from 15 to 18 metres. Fifteen metres is the wavelength most commonly used in the Samuels method, corresponding to twenty megacycles frequency.

2. *Effect of Treatment in Restoring Hormonal Balance*

When short-wave treatment is applied three times per week in accordance with the foregoing principles and with the proper dosage, spectroscopic analysis of the blood proves that extra-cellular hormonal equilibrium (equilibrium of hormone secretions in the blood and lymph) is attained after about three months of treatment, followed by intra-cellular equilibrium (proper functioning of the individual cells themselves) in the fourth month. Attainment of this equilibrium has a most profound effect on the general condition of the patient, and upon the symptoms which had resulted from glandular imbalance. Long before equilibrium is attained, the patients report a pronounced feeling of well-being from the treatments.

3. *Results of Treatment*

(a) *In Ailments Other than Cancer.* In diabetes, "normalization of the pituitary by short-wave irradiation results in a steady increase in production of insulin allowing an equivalent reduction in the dosage of exogenous (external) insulin, provided treatment is begun sufficiently early—that is to say, before irreversible damage has been caused to the greater part of the insulin-producing B cells of the pancreas. (This) treatment prevents any further deterioration of the insulin-producing cells."—Quotation is from Dr. Francis de Caux, op. cited.

Dr. de Caux makes the further point that in an established case already on a diet and insulin, treatment should begin at the earliest possible moment and the diet should be increased and the insulin decreased as the B cells start to recover. But an early case should, as soon as diagnosed, be treated with the short-wave therapy and not by diet and insulin, since in early cases a complete restoration of normal glandular functioning can often be accomplished.

A similar situation exists with multiple sclerosis and arthritis—in early cases, restoration of normal functioning is accomplished, but where irreversible damage has been done to nerve tracks or joints, the Samuels therapy prevents further deterioration, but cannot restore what has been irrevocably lost.

Excellent results have been obtained with the Samuels therapy in treating stomach and duodenal ulcers, various endocrine disturbances, and in many heart ailments, indicating the endocrine influence on the heart is greater than had formerly been recognized.

Many case histories are reported in the two books available on the subject of the Samuels therapy.

(b) *In Cancer, including Carcinoma, Sarcoma, Hodgkin's Disease.* In first and second stages of cancer, the Samuels treatment, as reported by both Dr. Samuels and Dr. de Caux, results in arresting the growth of the tumour, in most cases within three to four months. In one series of 150 cases, only 7 required more than four months to stop further growth of the tumour. The general condition of the patient responds soon after the start of treatment, and a condition of good health is regained, even in the few cases that require more than four months to stabilize the growth of the tumour.

In subsequent months, the tumour may shrink somewhat, but the main effects of the Samuels treatment are the systemic improvement, the avoidance of metasteses (spread of the cancer to other parts of the body) and cessation of further growth of the tumour.

Dr. de Caux states in his book already mentioned "The scalpel only removes the evidence and neglects the cause, whereas the short wave leaves the tumour but removes the cause, thus preventing further growth and extension of metasteses. As a consequence the patient has life and health restored without being mutilated in the process.

Dr. de Caux further states that with orthodox treatment, favourable results are obtained in about 50% of the second-stage cancer cases, while with the Samuels treatment, all patients starting treatment not later than the second stage are saved.

In third stage cancer, Dr. Samuels formerly saved 50% of the cases, which was 50% more than could be saved by orthodox treatment. Now, Dr. de Caux states that, with more intensive administration of the short-wave treatment (given for third-stage cancer cases, three times a week for at least a year), he is saving 80% of those cases. In third-stage cancer where extensive metasteses have occurred before the Samuels treatment is begun, the short-wave therapy is combined with large doses of testosterone.

4. *Advantage of Endogenous (Samuels short-wave) therapy over Exogenous Use of Glandular Extracts*

The aim of the Samuels endogenous therapy is to induce the patient's own endocrine glands to increase or decrease their own hormones as required. In general, Dr. Samuels is opposed to the administration of exogenous hormones—namely any hormones extracted from animal or synthetic sources and artificially administered to the patient, orally or by injection. He feels they merely fill the gap but do nothing to restore normal hormone production. However, in some cases it is necessary to use gland extracts to supplement the Samuels therapy.

Insulin is considered essential for established cases of diabetes, but the experience with the Samuels therapy shows that with pituitary irradiation, the dosage of insulin required can often be considerably reduced. Early cases of diabetes are stated as being curable without insulin.

When treating the climacteric, administration of small doses of the follicle hormone for a limited period, have been found of value when used simultaneously with the Samuels short-wave therapy.

The supplementary use of testosterone in third-stage cancer cases has already been mentioned.

With the exception of the above and similar situations, it has been found that endogenous endocrinopathy alone, is sufficient to restore the hormone-producing cells in the body to their normal activity. In this way the disease is cured, or, if it was in a very advanced stage before treatment, the patient's condition is ameliorated.

With exogenous hormones one rarely, if ever succeeds in bringing into and maintaining in equilibrium, the malfunctioning pituitary, state Drs. de Caux and Samuels. They emphasize that it is essential to restore the pituitary gland to normal in order to effect a permanent cure. They further list the disadvantages of exogenous glandular therapy as follows:

(a) When hormones are introduced into the body from the outside, the specific hormone-producing cells of the under-active gland gradually cease to function, and undergo atrophy from disuse. There is an additional decrease in the hormone being secreted, resulting in a greater need for hormones from the external sources, and thus a vicious circle arises, with increasing decline of internal hormone secretion. The result is that in the long run, exogenous glandular therapy often worsens rather than improves the condition of the patient's endocrine glands.

(b) The exogenous preparation is a poor substitute for one's own internal hormones, as glandular extracts from non-human sources are poor imitators of the human hormones. Also, administration of sizeable quantities of glandular extracts at intervals, does not have the same beneficial effect as the secretion of internal hormones which should take place constantly, second by second—day and night. Also, synthetic hormones cannot be expected to have the same effect as one's own natural secretions.

(c) The most effective dosage is not always known, and also the artificial hormone sometimes becomes inactive in the body, or may be eliminated without sufficient effect. If the dose given is insufficient, the desired result is entirely or partly lost.

In the case of an overdose, which cannot always be avoided, the endogenous hormone of the antagonistic gland may be raised, and this over-activity does not necessarily cease when the exogenous preparation is withdrawn. Also, excessive overdosage can cause hormonal damage in the body.

5. *The Effects of Biopsies and Surgery*

In very advanced cancer cases where surgery is deemed necessary as an emergency measure, it is advocated that the Samuels short-wave therapy be started immediately after the operation, in order to prevent the cancer cells liberated by surgery from forming either local recurrences or remote metasteses.

It is, of course, more desirable to start the Samuels treatment early enough so that no emergency surgery is required. The following quotation on the subjects of biopsy and surgery is taken from Dr. de Caux's book:

"Any interruption of continuity in the tissue of a malignant tumour, such as that caused by a biopsy, may cause the formation of remote metasteses before the time the patient arrives for treatment. Not always do these metasteses cause immediate complaints or symptoms. When it is remembered that not only the primary tumour, but also the metastes can continue growing so long as the pituitary hormones are not balanced, (that is to say during the first three or four months after starting the Samuels treatment) it can be realized that the possibility exists of these metasteses, especially those in the liver, having an unfavourable influence on the outcome of the treatment. . . .

"G. W. Taylor and R. H. Wallace, writing in the *New England Medical Journal* (Sept. 25, 1947, p. 475) stated that in reviewing the treatment of 382 patients with mammary cancer, of whom 236 had undergone an operation, it appeared that there were twice as many relapses in the cases previously sectioned as in the others. These authors concluded: 'Biopsy seems not entirely without hazard, since in those biopsied, recurrence in the operative field was twice as frequent.'

"The logical and common-sense procedure now is, therefore, instead of having a biopsy performed, to send the patient for a spectroscopic examination on the first suspicious symptom, and to initiate the Samuels therapy immediately (after) the spectroscopic diagnosis has been made. . . .

"If the tumour has already manifested itself, then immediate inception of short-wave therapy results in its growth being checked at an early stage and the patient thus saved from much misery and grief.

"Because in the early stages of development neither a radiogram (X-ray) nor any other method of examination short of a biopsy can establish a diagnosis, it is advisable to start endogenous endocrinopathy whenever there are suspicious symptoms. It is infinitely better—(to do this) than to perform an exploratory excision, curettement or puncture, which procedures only promote the formation of metasteses and so harm the patient. An exploratory laparotomy not only wastes time, but weakens the patient and lowers his powers of resistance and postpones unnecessarily the inception of the Samuels therapy.

"Samuels' advice is: no biopsy, but spectroscopical diagnosis; no operation, but immediate short-wave treatment."

6. The Role of Endogenous Endocrinopathy in Preventive Therapy.

Dr. Samuels has found through his research that all of the hormonal damaging diseases, including diabetes, cancer and arthritis, are preceded by a period of glandular imbalance before any outward symptoms begin. With the spectroscopic blood analysis, the glandular imbalance can be easily and quickly detected; and preventive treatment instituted early to balance the pituitary, thyroid, and gonads, would, according to the Samuels concepts and findings, prevent the development of any of the diseases traceable to hormonal imbalance. These diseases amount to over half of all illness, it is found. In this way, patients could be saved a great deal of discomfort and distress, and permanent damage to hormone-secreting cells and other body cells would be avoided.

IV. THE VIEWPOINT OF MEDICAL AUTHORITIES TOWARDS THE SAMUELS' WORK

Several years ago Dr. Samuels asked the Dutch government to appoint a commission to investigate the efficiency of his therapy. When the commission was appointed, it was headed by a Dr. Brutel, a former general practitioner who had never made a special study of either endocrinology or cancer, and who had for many years occupied an administrative post. Instead of choosing unprejudiced observers, Dr. Brutel filled the committee with five of Dr. Samuels' bitter opponents——

Dr. Brutel refused all contact with Dr. Samuels, and refused to examine any of the 130 patients that had been cured at that time. Instead, an adverse report was written, based on "unverified gossip"—and this report was widely circulated in pamphlet form in various countries. More recently, the continued success of the Samuels therapy with larger numbers of cases has forced the appointment of another investigation committee in Holland, and it remains to be seen whether official medical approval of the Samuels work will come now, or later.

Attempts have been made both in Holland and in England to stop the use of the Samuels therapy. In England, the Secretary to the British Empire Cancer Campaign claimed that the Samuels work had been "thoroughly investigated" by it, and as a result of its investigations, the Campaign decided that "no action should be taken by it in the matter". In reply, Dr. Samuels stated that no representative of the British Empire Cancer Campaign had come to see him up to that time.

The misrepresentation and bad faith evidenced by the above instances are unfortunately only too typical of the difficulties which a medical doctor faces in trying to gain recognition for new concepts or therapies which prove more successful in practice than the methods that are generally accepted. ...

Dr. Samuels' Policies

Dr. Samuels has spared no pains to make known his discoveries, and he welcomes observers at his clinic. He makes no charge whatever for either instruction or demonstration to physicians.

His first instruction course was held in 1938, the second in 1939, and after the war-time interval, others followed in 1947, 1948, 1949 and 1950. Doctors from many parts of the world attended these courses, witnessed various demonstrations presenting cases of successfully treated, and returned to their home countries, many of them to practice the Samuels therapy as a result of being convinced of its merits through personal observation.

Numerous case histories appear in the books cited on the subject, also references to professional journals carrying research articles supporting many of the conclusions summarized here.

THE "CAUSE" OF CANCER

Dr. Samuels and his advocates are jubilant over the discovery of what they consider to be the "cause" of cancer—namely the imbalance in pituitary function. To us, it rather seems that they have found a very important *link* in the chain of events which *induces* the appearance of cancer, and that the prime importance of the Samuels work is two-fold—partly in adding to our understanding of the cancer mechanism, and partly in providing one of the most effective treatment methods known for this devastating ailment.

There are two reasons, however, why we cannot consider the Samuels concepts as the complete solution to the problem of the "cause" of cancer. One, is the fact that they do not present any convincing evidence regarding the reasons why the pituitary functions become imbalanced. They discuss hereditary and acquired factors, but in that respect their discussion becomes vague, in contrast to the precision of the rest of their presentation. Secondly, there are many cases in which severe pituitary malfunction does *not* lead to cancer, but instead to other ailments. Here, Dr. Samuels and his followers fall back on the rather indefinite idea of "pre-disposition to tumour"—if a patient has this pre-disposition, so runs the Samuels reasoning, then pituitary imbalance produces cancer. "Pre-disposition" is also too vague, when we have the far more precise means of analysis provided by R. R. Rife and Dr. Wilhelm Reich, for delineating the specific conditions under which tumours develop in the body.

However, this should not be construed as a criticism of the Samuels work. The problem of cancer is probably far too large for any one individual to obtain a complete grasp of the entire subject, and it appears that there is no one single cause, but rather a chain of many factors. There is a tendency for each investigator and discoverer in this field to regard his own work as the complete answer, whereas it would seem nearer the actual facts to accord recognition to various brilliant scientists for their respective contributions to the subject.

R. R. Rife demonstrated the importance of slight chemical changes in starting a cycle by which the common colon bacillus goes through several stages and eventually ends as a cancer-producing micro-organism. In this regard, it should be kept in mind that the pituitary plays a most important role in regulating body chemistry; there need be no conflict between the Rife and Samuels theories—it would seem that they complement each other.

Dr. Wilhelm Reich's extensive research disclosed many new facts about the precise mechanism by which tumours are initiated, grow, and how they can later poison the entire bodily mechanism. He also has presented much material on the effect of mental and emotional inhibitions in interfering with the flow of fundamental energies which normally maintain health in the organism. It would seem that both Rife and Reich's work have a very appropriate place in supplying the gaps in the Samuels concepts, and vice versa. Since mental and emotional disturbances are known to have a very profound effect on glandular functioning, the inter-action between the Reich orgone concepts and the Samuels concepts becomes evident. Each has a part of the picture—what is needed is an organization which will take all parts and fit them together. It is indeed un-unfortunate that the cancer research organizations which solicit large sums of money from the public, have only contributed to the support of certain factions of medical research, completely ignoring the work of the groups which have had actual and outstanding success in treating cancer, and likewise ignoring those who have accomplished a great deal towards a better understanding of the fundamental factors involved in the production of cancer.

It is interesting to note that those who discover the greatest detail about the different aspects of the cancer process, are not necessarily the ones who develop the most effective treatment methods. For example, Dr. Wilhelm Reich discovered far more detail regarding the development of the cancer process than did Dr. Samuels, yet Reich is not too successful in treating third-stage cancer, judging from the case histories given in his book *The Cancer Biopathy*, whereas Dr. Samuels and his followers obtain much better results with their method of treatment applied to third-stage cancer. Likewise, Lakhovsky's treatment method produces outstanding successes in third-stage cancer, though his contributions to the understanding of the cancer process are not as extensive as those of Reich.

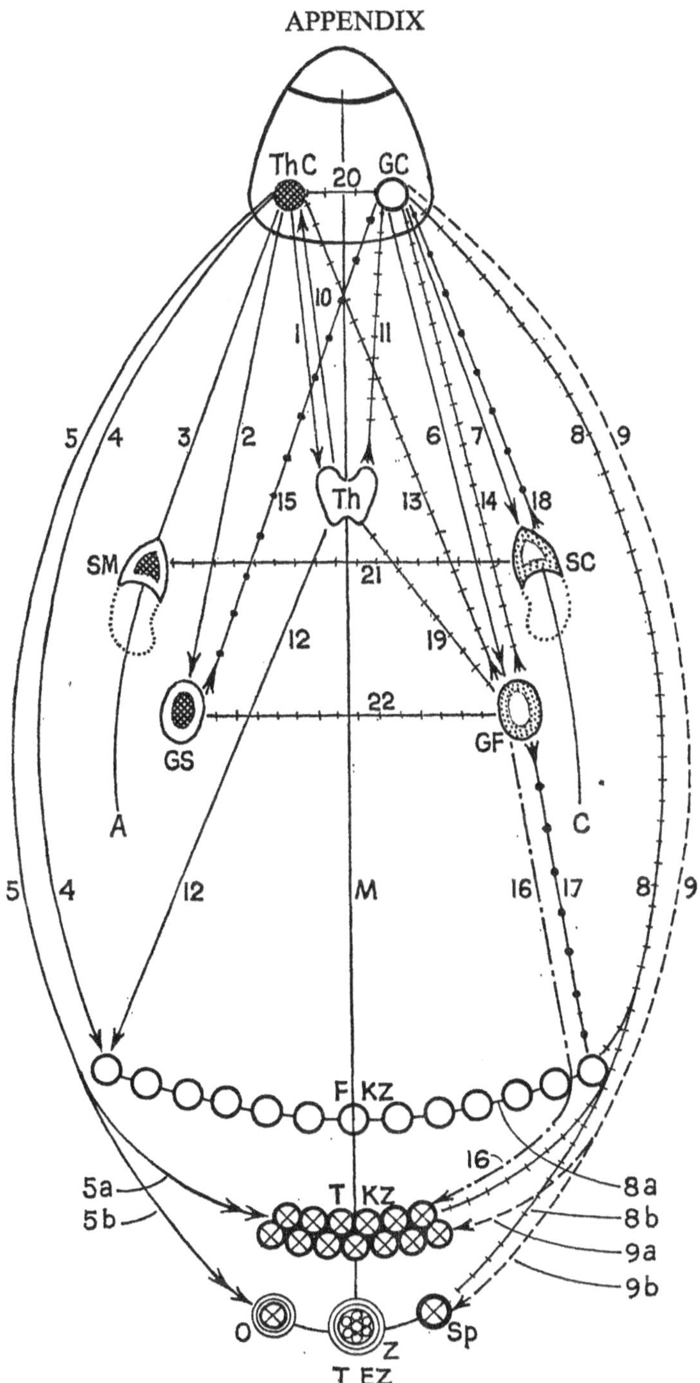

Schematic figure showing correlation of the principle hormones

247

The correlation between the two pituitary hormones and the hormones of the thyroid, sex-glands and suprarenal glands, and the statistical representation showing how the two hormones influence in a regulating manner the cell functions and the cell dividing process.

ThC   = thyrotropic centre of the pituitary.

GC   = gonadotropic centre of the pituitary.

Th   = thyroid.

GS   = testosterone or luteo-hormone production of the sex glands.

GF   = follicular hormone production of the sex glands.

SM   = adrenalin production in the medulla of the suprarenal glands.

SC   = cortin production in the cortex of the suprarenal glands.

FKZ   = functions of the body cells.

TKZ   = the dividing process of the body cells.

TEZ   = dividing process in the female and male gametocytes, in the zygote and of embryonic or foetal cells.

Divisions of the:

O   = ovocytes or female germ cells.

Sp   = spermatocytes or male germ cells.

Z   = zygote, embryonic or foetal cells.

⟶   = direction of the stimulating action of certain glands.

   = direction of the checking or inhibiting action of certain glands.

   = strongly checking action of certain glands.

   = checking action of the second degree.

   = the cell- and egg cell-activating action of the gonadotropic hormone.

   = the stimulating action of the thyrotropic hormone on the cell- and egg cell-dividing-process.

—·——·—   = the action of the follicle hormone as proliferation hormone of the second degree.

    A = the adrenalin secretion in the blood.

    C = the cortin secretion in the blood.

   M = median line.

1. The thyrotropic centre stimulates the hormone production of the thyroid.

2. The thyrotropic hormone stimulates the testosterone- or the luteo-hormone production of the sex glands.

3. The thyrotropic hormone stimulates the adrenalin production of the supra-renal glands.

4. The thyrotropic hormone is the stimulating factor of the cell functions.

5. The thyrotropic hormone is the stimulating factor of the cell dividing process (5a) and of the egg dividing process (5b).

6. The gonadotropic hormone is the activator of the follicle hormone pro-duction of the sex glands.

7. The gonadotropic hormone is the activator of the cortin production of the suprarenal glands.

8. The gonadotropic hormone is the checking factor of the cell functions, of the cell dividing process (8a) and the egg cell dividing process (8b).

9. The gonadotropic hormone is the activator of the cell dividing process (9a) and the egg cell dividing process (9b).

10. The thyroid hormone is the synergist of the thyrotropic hormone.

11. The thyroid hormone is the antagonist of the gonadotropic hormone.

12. the thyroid hormone is the stimulating factor of the cell functions and as such the synergist of the thyrotropic hormone.

13. The hormones of the sex glands check the thyrotropic centre.

14. The follicle hormone keeps the gonadotropic centre balanced and is there-fore as such the antagonist of the gonadotropic hormone.

15. The luteo hormone or testosterone checks the gonadotropic centre, but to a lesser extent than the follicle hormone.

16. The follicle hormone is the second proliferation hormone.

17. The follicle hormone exercises a checking action on the cell functions and as such is the synergist of the gonadotropic hormone.

18. The cortin is a checking factor of the second degree of the gonadotropic centre.

19. The thyroid hormone and the follicle hormone are each other's antagonists.

20. The thyrotropic hormone and the gonadotropic hormone are each other's antagonists and therefore keep each other balanced.

21. The adrenalin is the antagonist of the cortin; both hormones keep each other balanced.

22. The follicle hormone and the luteo hormone or testosterone keep each other balanced and as such are each other's antagonists.

EXCERPTS FROM

*RADIATIONS AND WAVES — SOURCE OF OUR LIFE**

by GEORGES LAKHOVSKY – 1941

"Before publishing the therapeutic results obtained with my multiple short wave oscillator and presented at the International Congress of Short Waves in Vienna (July, 1937) it may be useful to give a rapid survey of my theory of cellular oscillation which I have developed fully in a number of books (*The Secret of Life; Contribution to the Etiology of Cancer; The Earth and Ourselves; Cellular Oscillation; Nature and her Wonders.*)

All living cells are composed of two essential elements; the nucleus and the protoplasm in which it is bathed. This nucleus is itself composed of many tubular filaments: the chromosomes. In addition, hundreds of such smaller filaments or chondromes are present in the cytoplasm.

Chromosomes and chondromes are sheathed in an insulating substance (cholesterine, resin, fat, plastrin, etc.) and contain a liquid like serum with the same mineral content as sea water, and consequently a conductor of electricity. Thus, these filaments constitute ultramicroscopic oscillating circuits capable of oscillating electrically over a wide scale of very short wave lengths.

I have demonstrated in my works that these cellular oscillating circuits, chromosomes and chondromes, vibrate electrically under the stimulus of electromagnetic waves: cosmic, atmospheric and telluric.

Now, many internal and external influences may upset the oscillating equilibrium of these cells. For instance, a variation or change in the field of cosmic, telluric or atmospheric waves, a demineralization of the organic matter constituting the cellular substance, traumas causing the destruction by shock of the protoplasm or the nucleus.

I have shown in my books, *The Secret of Life*, and especially in *The Earth and Ourselves*, that every living cell draws its oscillatory energy from the field of secondary radiations resulting from the ionization of the geological substances of the earth by cosmic radiations.

But certain natural radiations are particularly toxic, especially those originating in earth grograms. Many cancer cases have been attributed to these toxic radiations and proven experimentally, notably in Germany by Dr. Rambeau of Marburg. Therefore, earth radiations sometimes cause disturbance of the cellular oscillatory equilibrium of the organism.

*Published by Emile L. Cabella, 228 East 45th St., New York City, USA.

Under these varied circumstances cellular oscillation may cease; the cell is then dead. But within the dead cell, the chondromes sometimes continue to oscillate electrically on their own natural frequencies. Fortunately, this phenomenon occurs rarely, or all mankind would already have perished from cancer.

The chondromes then envelope themselves in a membrane and continue to oscillate and multiply independently of the cell. They may then become neoplasic cells.

To re-establish this equilibrium, I thought of creating, in 1923, a constant compensating field of very short radiations (2 to 10 metres in wave lengths) to neutralize the action of the disturbing rays, and give the living cell the necessary stimulation for a return to its normal oscillation.

To this end, in 1923, I constructed my short-wave oscillator, using two triode tubes for very short waves made especially for this apparatus. I tried several cross leakages for this machine using one or more tubes and then multiple triodes with a tube containing oscillatory circuits within the bulb. Finally I adopted the oscillator with symmetrical cross leakage comprising two triodes. The oscillating circuits formed a single spiral, branched between the two grilles and the two anodes. It was fed directly by alternating current from the local supply circuit.

With this short-wave apparatus I was able to cure plants inoculated with cancer. For six years at the Salpetriere I observed and checked the effects of these short waves.

Using very low power, from 10 to 12 watts, and a limited duration of treatment, I succeeded in curing cancer in human beings, but also had to record some failures.

The news of the success of my experiments became widespread. In many countries, as early as 1928, they began to build short wave oscillators of considerable power producing thermal effects.

But there was great danger that the chromosomes and chondromes, which are barely a ten-thousandth or twenty-thousandth of a millimeter in thickness, might not survive under a high frequency current. They offer much resistance, even to a low current which is sufficient to dissolve and destroy them.

It is simple to prove this by bringing a small bulb of from 2 to 5 volts with a filament of several hundredths of a millimeter, inserted between metal rods forming antennae, within the radius of a short wave transmitter. The bulb will light up and sometimes burn out, if it is brought too near the apparatus.

Moreover, the chondromes and chromosomes of all living cells, which are infinitely finer than the filaments of the bulbs, are sensitive centres of thermal phenomena, which may provoke their fusion. Undoubtedly this method is effective in killing microbes in the organism and in neoplasic cells. But it can also destroy millions of cells of healthy tissue in every irradiation.

On the other hand, I thought it possible to obtain better results by administering an oscillatory shock to all the cells in the organism simultaneously. Such a shock, very brief, produced by damped or weakened electrostatic waves, does not provoke thermal and prolonged effects and involves no risk of burning living cells.

I therefore sought to produce an artificial oscillatory shock causing a periodic oscillation of the weak or dead cells.

At first glance this problem seems physically insoluble as there are approximately 200 quintillion cells in our bodies, each oscillating on its own natural wave length. Theoretically, therefore, we would have to have a different wave length for each cell, so that every cell in the organism would oscillate in resonance on its own wave length.

After much research I was able to construct an apparatus creating an electrostatic field covering all frequencies from 3 metres to the infra-red, so that every cell can find its natural frequency and vibrate in resonance.

We know that in physics, a circuit fed by damped or weakened high frequency currents creates many harmonics. Consequently, I conceived an oscillator of multiple wave lengths with a broad scale in which every organ, every gland, every tissue, every nerve, could find its natural frequency.

To obtain this result I set up a transmitter composed of a series of circular concentric oscillating circuits linked by a silk cord but not contiguous. These circuits are stimulated by damped high frequency currents from a spark gap. Thus each circuit of the transmitter vibrates not only on its natural frequency, but also on numerous harmonics.

Thus, I built an oscillator with all the basic wave lengths from 10 centimetres to 400 metres, that is, all frequencies from 750,000 per second to 3 billion. But each circuit also emits many harmonics, which, with their basic waves, their interferences and their effluvia can reach the scale of infra-red and even that of visible light (1 to 300 trillion vibrations per second).

Since all the cells as well as the chondromes oscillate precisely at frequencies on those gamuts, they can therefore find on the scale of such an oscillator, the frequencies which cause them to vibrate in resonance.

You know the results I obtained with continuous very short wave using triode tubes at a distance, with no contact electrode. . . .

As early as 1931, I began using this multiple wave oscillator in various Paris hospitals: the Saint Louis, Val-de-Grace, Calvary, Necker, etc. . . . Among the many cures with this treatment, I will mention especially those of various cases of cancer which X-ray and radium treatments failed to improve. These patients cured six years ago, have had no recurrence and are in perfect health at this date. In all pathogenic cases this treatment gives very good results. As it does not attack the microbes directly, it does not destroy live tissue, but reinforces the vitality of the organism by accelerating cellular oscillation. It is therefore the reinforced organism that successfully resists the microbes and all pathogenic causes.

So, while X-rays and radium destroy microbes, neoplasic cells and healthy tissues at the same time—which accounts for the serious accidents which occur during and after such treatments—high frequency radiations (short waves) applied at a distance and without thermal effect cure diseases of all kinds, even those of the prostate to a considerable degree.

Whatever the pathogenic cause, the multiple wave oscillator reproduces the frequencies necessary to re-establish the cellular oscillatory equilibrium.

In general, it is sufficient to seat the patient, or to have him stand in the radius of the apparatus, before the transmitter. The duration of the treatment is usually from five to ten minutes, every other day. These figures are purely arbitrary, since these radiations reinforcing cellular oscillation do not produce organic disturbances, whatever the duration of the treatment may be.

Despite the many cases successfully treated, almost without exception, yet it must be understood that my oscillator cannot cure all types of cancer in all its stages of development. In many cases, when the cancerous tumour has already destroyed important blood vessels, my apparatus is powerless to re-build that tissue before the fatal haemorrhage."

THE MAGNETIC-ARC INSTRUMENT OF S. S. KNIGHT

Mr. Knight, a chemical and electrical engineer of Walnut Creek, California, was impressed with the severity and persistence of pain suffered in certain types of cancer. Realizing that pain is a sensation representative of an electrical phenomenon of excessive nerve-current flow, he decided to experiment to see if there was some way of interfering with that nerve flow through artificial impulses, in the effort to stop or alleviate the pain.

Towards this end, he developed an instrument combining several features of electro-therapy:

1. A wave pulsed at 90 times per minute, taking advantage of data obtained from experience with the older types of Faradic machines.

2. A rest interval exceeding 75% of the duration of each pulsing cycle—enhances the beneficial effect.

3. Breaking the treatment by mechanical opening of contacts in such a way as to create an arc at each of the pulsing intervals. An arc produces a very wide band of frequencies, from those of very low frequency, to the highest that can be measured. The multiplicity of frequencies seems to play a part in masking or interrupting the sensation of pain.

4. A unique feature of the Knight instrument is that the arc is broken in a strong magnetic field produced by a large, powerful, permanent magnet. This introduces what is called the "Zeeman effect", that is, a magnification of the arc impulses, and a spreading of frequencies.

5. Use of very low power output—below 1/100 of a watt, so that there will be no perceptible heating of tissue when the complex of impulses from the instrument are fed to the patient through electrodes applied to the skin.

Experiments on cases which had been declared terminal cancer by medical doctors, showed that this device not only brought outstanding relief of pain in the majority of instances, but also produced some surprising "remissions" characterized by shrinking of tumour masses, and restoration of general health to the patient.

Dr. C. P. Bryant of Seattle, Washington, who used these instruments extensively in his practice over a period of years, declared that with their use he was able to restore to health 75% of all cancer cases which had not undergone surgery, X-ray or radium, and 25% of those cancer cases which had been subjected to surgery, X-ray or radium.

The Knight family has never represented these instruments as a "cancer cure". The original aim was to find some way to relieve the pain of cancer. Reports from doctors of apparent cures through use of the instruments were received with interest, as well as reports of benefits from the application of the instrument to conditions other than cancer.

Mr. S. S. Knight, a man with pronounced humanitarian ideals, made the circuit of the instrument available to all inquirers and encouraged the building of instruments by anyone who wished to copy his circuit and specifications. He was keenly disappointed when reports reached him of failure, or lack of therapeutic effect from instruments constructed by others. Upon examining some of these devices, he then realized that others did not have his knack of adjusting the instrument for maximum effect. It developed that adjustments were quite critical, particularly concerning the making and breaking of the arc. Mr. Knight could tell by looking at the arc and listening to it, when it was "right". It was difficult to develop objective criteria for adjustment of the equipment.

A puzzling circumstance, was that the maximum effect from Mr. Knight's instruments was obtained when they were new; after a few weeks of use the effect began to diminish; after some months it was perceptibly weaker, and after a year or more, of very little use. Several years ago, the reason for the lessening of effect was determined. The isolation transformer used to block the direct flow of the arc frequencies from the electric line supply (an isolation required to eliminate radio interference) was being saturated by the magnetism from the large magnet surrounding the arc. The more saturated the transformer became, the less electricity it passed. The equipment was re-designed to place the transformer out of range of the field of the magnet.

The instrument can be classed with experimental devices, as there have not been sufficient tests made under controlled conditions to establish with certainty the percentage of cases of various types of cancer that can be alleviated, arrested, and perhaps cured, but the tests to date indicate at least that this is a promising approach, which should be explored further.

An imitation instrument has been marketed by others, using an electromagnet instead of a permanent magnet. Mr. Knight was opposed to this substitution, having a number of reasons for believing that the permanent magnet was preferable. Other changes have also been made in constructing these imitation instruments, including the type of electrodes. So far as your compiler has been able to determine, the instruments using electro-magnets have some value in relieving pain in a variety of conditions, but the effect on cancer seems to be slower and less pronounced than obtains from the genuine Knight instruments when the latter are functioning properly and are in adjustment.